RUN FAST

THE DEFINITIVE GUIDE TO ACCELERATING TECHNOLOGY PROJECTS

NEIL HOW

RETHINK PRESS

First published in Great Britain 2018
by Rethink Press (www.rethinkpress.com)

Contents

Introduction

No matter what industry or company you work in, everything is based on projects. Whether it's a new product, a marketing campaign, a multimillion-pound acquisition or the office party, the desired and arguably best management structure is some form of a temporary project. Whether we like it or not, projects are everywhere.

It's probably fair to say technology-based projects have not had the best reputation over the years. These projects, for whatever reason, have always been problematic. You can always find a case study of a technology project that went wrong. The degree of wrong is variable, but ask the fundamental questions of time, cost and scope, and inevitably one of those factors will have been a major issue. From the public sector to oil and gas, banking or retail, technology projects always seem to be a challenge. Over the last 20 years or so the project industry responded, and now we have many highly robust delivery models – Prince2, PMP, Agile, Activate/ASAP and other software-specific deployment methodologies. These are a huge asset to anyone driving a project, but despite these, the fundamental problem remains: why are projects still failing and proving to be such exasperating work?

In the world of project delivery, when the delivery dust has settled, there are three types of projects; those that are effortless, those that are infuriating and those that fail.

- The effortless projects sail through design and delivery, come in under budget and deliver ahead of the planned timescales. They are the Rolls Royce of projects, championed by a smiling project manager who dances through organisational politics, builds long-standing and deep relationships with everyone involved, and whose primary challenge is fending off colleagues fighting to be on their team. Nothing is too big for this team. They are the rainmakers of delivery and receive praise once the project is live.

- The next group along the scale are the projects that require pure hard work. These are the 1970s, smoke-billowing Vauxhall Vivas. They typically run for months and months, they cost the earth, run everyone into the ground and drag themselves over the line while everyone around them collapses with exhaustion. The business change is resisted, tasks are barely completed and the impact on the business is felt for years. They achieve the required outcome, but the cost to the organisation is significant – in terms of monetary value, time, opportunity cost and emotional baggage. Everyone is glad to get these projects over the line so all involved can sit in a dark room and recover from the experience.

- The last group is those that fail. These are the Reliant Robins. They start with good intentions but quickly deteriorate into carnage and confusion. They are poorly executed, badly governed, not fit for purpose and are seen as a burden to the organisation. The delivery process causes so much pain and upset to an organisation that morale hits rock-bottom. They seldom get over the finish line and are often cancelled midway through delivery as the organisation seeks a damage limitation approach.

Regardless of the company, everyone – whether a senior executive, a project leader or a technical supplier – aspires to the Rolls Royce project. However, for one reason or another, the technology industry has struggled to achieve this. Even with some of the best people at the helm, many projects are inherently difficult and seldom achieve this status.

And then, as if things were not complicated enough, there is a new, external, challenge. The world is changing around us. We are now in what analysts are calling the 'digital era'. This brings with it obstacles that no other era has faced. Companies are changing and evolving at a pace never seen before, markets are being disrupted by innovative start-ups, and two new generations of workers are now entering and driving the workforce. Expectations are immediate, costs tight and timescales ever shorter. The modern digital business only plans six months ahead, as anything else is subject to too much volatility. With this in mind, projects can only logically have a duration of six months! Chief executives around the globe cite a frustration that timescales on technology projects are still too long and costs too high. They are looking for a better approach.

The answer, quite obviously, is that we need to get better at projects. We also need to get better at doing so within the era we find ourselves. The digital requirements placed upon those involved in delivery have changed, and cost and timescales have never been more important.

This book looks at the key themes used to accelerate technology projects, ultimately to achieve a maximum six-month project. Anything longer is not viable as the business case will likely have expired. Six principles, grouped together as the PROMPT method, will be introduced and explained with examples and actionable checklists. These principles are drawn from speaking with technology professionals around the globe, analysing projects that went well (and those that didn't), research papers from academics and analysts, and countless years of experience. The themes are broad, but when pieced together create a culture and foundation that can supercharge a project, delivering it in a short timeframe with robust governance. The knock-on effect is a reduction in overall cost, as typically the highest cost to a project is people. If cost reduces, then the return on investment (ROI) increases and the payback period naturally decreases. Good news for everyone.

What makes me qualified to write this book? Well, this book is not about me or my business, but about you and yours. Over the last twenty years, I have had the pleasure of running some extraordinary projects – from small-scale bespoke software developments to multimillion-pound global enterprise resource planning (ERP) implementations. I used to look at my industry and peers and think myself fortunate. Most of my projects were not painful – I actually found them rather enjoyable. I would probably go so far as to say I love what I do (my wife thinks I'm crazy!). Of course, certain aspects were tough at times – but they seldom went wrong or failed to give the customer the outcome they desired. People enjoyed working with me – customers and team members alike. We had a good project social life and, indeed, many of my colleagues are now family friends. Happy days.

As my experience and reputation grew I met more and more people through my network. At this point I started to notice the industry reputation on a macro level.

There were lots of complaints from disgruntled chief finance officers (CFOs) about the cost of failing projects. Some of the chief information officers (CIOs) or chief technology officers (CTOs) and technology leaders felt threatened by projects that were going wrong. One CFO broke down in tears while I was talking to him, worried about losing his job and the impact the project was having on his family life. Stress seemed to feature heavily in all conversations and felt so tangible it was almost part of the job description. I struggled to understand why it seemed so difficult for everyone else. Why was my experience so different?

I set up my company five years ago with a mission to run large technology projects using the same strategies and principles I had successfully used for years. I wanted to dispel the stigma in the marketplace and show everyone it didn't have to be difficult to run great projects. I researched the themes central to this book and created a methodology based on everything I had discovered – business was good. Then I hit a problem: I was being asked to serve and help more and more people. More than I could possibly accommodate, even with my team. I needed a new approach.

This book was put together to serve the greater market. It's for all the people I have not had the pleasure of working with directly. I have tried to unpack the contents of my head, the soft and technical skills, and explain the methods I have used throughout my career.

I truly hope this book helps you. Even if you can't implement all of the steps, anything you adopt will drastically improve your chance of success. I implore you to read and bring these principles into your project teams. Make sure your leadership teams understand them and abide by them. Start driving success from within – it's the only way. Whether we like it or not, technology is now a strategic force in the digital era. Let's do ourselves and our companies proud, deliver as we are required to and dispel the tarnished reputation this industry has.

Good luck!

Neil

CHAPTER ONE
The Legacy

The history of technology delivery

We must understand our failures to guard against them happening again, and for that we have to start at the beginning. The past protects us from the same mistakes – assuming we actually change something.

> 'Life can only be understood backwards; but it must be lived forwards.'
> —Soren Kierkegaard

The delivery of technology has its roots in the Industrial Revolution, from 1760 to 1840. At that time, wealthy business owners recognised and implemented the growing trend of steam-powered machinery to speed up their industry processes (typically manufacturing). Mills were automated through steam-powered looms, mines dug deeper using water pumps, and both infrastructure and agriculture improved dramatically.

The second industrial revolution, often called the technological revolution, took place from 1870 to 1914. This was centred predominantly around further enhancements in manufacturing, including interchangeable tooling and the introduction of new materials (steel, iron, etc). These improvements coincided with the introduction of railroads, telegrams, the electrification of cities, the telephone and the production line. The movement of people began, and ideas and innovations were shared globally and exploited readily.

Then the world went through two world wars. This era of innovation was more focused on the war effort – weaponry and aircraft (tanks, machine guns, radar), medicine (portable X-rays, sterilisation, bandages), and food production (fertiliser).

Post-World War Two, the world went through an economic boom. Known as the golden age of capitalism, this era experienced unusually high and sustained growth. Not a single industry failed to prosper during this era. A 1957 speech by British Prime Minister Harold Macmillan captures what the golden age felt like:

> 'Let us be frank about it: most of our people have never had it so good. Go around the country, go

> to the industrial towns, go to the farms and you will see a state of prosperity such as we have never had in my lifetime – nor indeed in the history of this country.'[1]

To list the innovations during this period is a book in itself, but the era is defined by three technical ages: the atomic age, the jet age and the space age. Each age stepped us through countless inventions.

The golden age was brought to a close by a number of events coinciding in the 1970s: the collapse of the monetary system in 1971, the 1973 oil crisis, the 1973–74 stock market crash, and finally the 1973–75 recession.

Then we reach the information or digital era. This era marks the start of technology as we know it today. Analogue devices are being replaced by their digital equivalent, fibre optics and microprocessors continue to push boundaries, and the internet continues to facilitate global trade and communications. Switches and sensors have become increasingly advanced and the personal and handheld computer (smartphone) is now in everyone's pocket. Throughout this era, the world and the objects within it became, and continue to become, increasingly connected.

What does this history lesson tell us, apart from highlighting the speed of evolution over the last 150 years?

1 BBC website: 'Britons "have never had it so good".' 20 July 1957, retrieved 12 March 2009.

We need to understand that we have always innovated and delivered incredible technology projects. The challenge we need to put to ourselves is this: why is the delivery of technology in the digital era so complicated compared to a project involving any of the innovations in the previous eras?

Change in technology across the various eras

To unpack the answer to this question we first need to understand the degrees of complexity across each of the eras. Fundamentally the level of complexity increases throughout each of the eras before it's reset to a new baseline in the next period. This is down to the combinatorial innovation that occurs throughout the period. People take the innovations of that era and build upon them, creating something new and desirable. Then at the end of the era, something occurs that changes the entire landscape: electrification replaces the steam engine, digital replaces analogue. At that point, all previous innovations are quickly made redundant, and the cycle begins again. We are currently forty years into the digital era, making it a highly mature era and, following the pattern, a highly complex era.

The second factor that led to a high degree of complexity in the digital era is speed. As the eras have progressed, the degree and speed of communications has provided many new opportunities. Less than 100 years ago, it would have been difficult to hear about a development in a foreign country. Today it involves a few taps on a smartphone. The speed of information sharing and access to news and knowledge has led to innovation and development at a rate we've never seen before. Today's technology is out of date before it has even been deployed in the workplace.

Finally, the era has seen exponential growth in the processing capabilities underpinning the new technology. The first mainstream IBM personal computer was created in 1981. It had 16KB of memory and no internal storage. Today, hardware is smaller, cheaper and has endless capacity. The majority of current innovation sits on top of a technology platform of some nature. The better the platform, the more capabilities that are unlocked. This exponential explosion of the technology underpinning innovation has led to more options, more simplification and even more combinatorial innovation.

The most prevalent example of this combinatorial innovation is perhaps the Amazon Alexa, the personal assistant who, with speech, can do almost anything you ask. Under the stylish veneer, the technology is listening to and processing human diction in real time. On command, it fires off the request to whatever application or third-party interface is required. Within milliseconds, a response is formulated and either the action performed or answer spoken back to the user. At the time of writing, there are thousands of integration points for Alexa, and a comprehensive software development kit to enable developers to build custom integrations. This device alone illustrates the complex world we now operate in.

Given this complexity, and the speed of innovation, our next question must be: does an increase in innovation complexity require an increase in the complexity of project delivery?

Immaturity of technology delivery

By our very nature, the human species is designed to reflect on our decisions and actions and develop a better process. How we deliver and execute will adapt and change through time. Fundamentally, it's how we have survived and evolved.

The challenge is that with each passing unit of time, the rate and pace of change has increased so rapidly that we are struggling to learn from our previous experience. What we implement today will be replaced by the thinking of tomorrow. New technology and the discrete trends it operates within require different thinking. The technicians among us are almost always on the back foot, fighting to keep abreast of the constant innovations.

Although the digital era is more than forty years old, we must remember that each technical step or innovation resets the thinking as to how things should be done.

In addition to technological advances, how we conduct business today continues to shift rapidly. We have different expectations and ideas, a different philosophy of management, and a different generation of workers with a different outlook on the world. As the expectations of management change, we must innovate how we deliver projects.

The continually shifting technology base, the speed of innovation, and the changing nature of how companies operate force the delivery of technology projects into an immature state. At best, we can only use the learnings from the past three to five years to inform how we deliver projects.

The industry has rallied around, and to its credit, worked hard to define and protect against these challenges. We have delivery frameworks such as Prince2, Agile and PMP, as well as some vendor-based frameworks such as SAP's ASAP and Activate. These frameworks provide us with a structured, industry-recognised, templated approach; however, one size does not fit all and the delivery mechanism needs to be adjusted and tweaked for each delivery.

Is it any wonder that technology projects go wrong?

What is failure?

The Project Management Institute, in its 2017 *Pulse of the Profession* report, suggests that only 69% of technology projects deliver their expected outcomes.[2] The same research finds that only 57% finish within budget and 51% finish within the timescale planned. Amazingly, 49% of projects experience scope creep or uncontrolled change. These, regardless of anyone's background, are quite shocking results. While you might consider some failure acceptable – business scope change, for example – the cost of these failures is still significant. If the project is designed and structured appropriately from the outset, the scope should not change or the project should be able to absorb it with no additional cost.

There is no way to spin the failure of technology projects. If they are started and managed correctly, they should be able to deliver against all three key metrics (cost, time, and quality/scope).

Reasons for failure

Given the shockingly high failure rates of technology projects, we need to look at the most common causes of these failures to protect ourselves from them in the future.

2 www.pmi.org/-/media/pmi/documents/public/pdf/learning/thought-leadership/pulse/pulse-of-the-profession-2017

Poorly defined outcome: An ill-defined outcome for a project is all too common. Typically, leaders are great at idea generation, but poor at explaining the measurable impact to the organisation. What does improved customer service mean in terms of success? Shorter call times? Fewer calls? Higher customer satisfaction? How will you know when you've succeeded? If you don't know, you're doomed to failure. The problem is magnified when other senior leaders are involved. Everyone has their own view of success. The net result is unclear goals, expanded scope and poor design, all of which impact delivery timelines and cost.

Lack of leadership: There is a trend for technology projects to be managed by technology people, regardless of what the outcome of the project actually is. If the project doesn't have an endorsement from senior level executives (typically called C-suite executives), it will be extremely difficult to get people on board and to determine who is responsible for making the tough decisions. Related to leadership is the skill of the project manager (PM). A poor-performing or unskilled PM will not understand the organisation and will not balance the needs of the project with the needs of solid governance. A lack of subject matter expertise can also hinder project managers. Like it or not, sometimes you have to get into the weeds.

Lack of accountability: When projects are dubbed 'technology projects' and left to the technology department to manage, it tends to struggle to gain commitment and buy-in from the business teams. This results in a lack of ownership and accountability, often leading to failure through poor change management as the business teams do not adopt the new way of operating. Typically, technical teams do not have wide enough accountability to resolve business problems.

Linked to this is the danger of having too many people. Large projects tend to have a variety of senior executives involved. Each executive will have a subtly different agenda and expectation of the project benefits and outcomes. At times, these are incompatible and need managing.

Weak sponsorship also indicates poor accountability across the project. When dates slip, tasks aren't completed or roadblocks emerge and no one is held accountable, the project quickly becomes unmanageable. Weak ownership often results in poor project control which results in an increased chance of the project failing.

Poor governance: It's a common assumption that lack of governance is a major reason for the failure of a technology project. Ironically, most large projects have the exact opposite problem: there is too much bureaucracy. In large organisations, stakeholders such as risk managers, compliance staff, methodology experts and architects all have their own governance demands, which greatly increases the demands on project staff. The right balance needs to be achieved.

Insufficient communication: Projects in the digital era are complicated organisations in themselves. Lack of communication, or misguided or incorrect messaging, can stop a project dead and cause it to idle for weeks.

No plan or timeline: Without a clear plan it's impossible to know if a project is on schedule. A plan also needs to be monitored and adapted, not just filed away when complete. It's a living and breathing document.

Solving the wrong problem: Companies may think they are doing something to address their problem, but in reality they are addressing the wrong problem. In our customer service example, if shorter call times are the only metric for improved customer service, employees will be encouraged to get off the phone quickly, which may or may not actually improve customer service. Call time decreases, but customers may be even less satisfied than before.

Over-optimism: We all want our projects to succeed, but in selling the project to others, do not underestimate the costs or over-emphasise the benefits. Business cases have a tendency to be exaggerated to achieve the required ROI. Sometimes this is not intentional but instead results from a lack of understanding of the situation and requirements.

Complexity: We have already established that the digital era is one of the most complicated environments – risks and effort increase exponentially as complexity increases. Systems become strained as technicians try to cater for this complexity in tight timescales, sometimes resorting to clumsy workarounds. The complexity overwhelms the project and it gets out of control.

Over-engineered projects: The complex world we are delivering projects in often creates over-engineered delivery vehicles. The recent trend of valuing form over substance needs to be addressed. A CIO recently said,

> 'In the past, we spent time working out how to solve a problem. We explored different avenues of approach, different ways of meeting a target. We focused on the customer's needs and expectations. We focused on what was to be delivered. Nowadays, it seems that the focus is more on planning how you are to do it, coupled with a determination to adopt methodologies and standards. The result is that we spend more time planning the methodology and approach to the project rather than working on the technical requirements of the solution and delivering it. The consequence is it gives rise to a much greater number of formally documented meetings, implying a new breed of project administrators who manage the documentation and schedule the meetings, adding to the project overhead, both in time and cost. A second consequence is that the poor people at the coal face have, in addition to actually doing the work, a welter of forms to complete.'

This section wouldn't be complete without a few case studies to illustrate how easily projects in the digital era go wrong.

NATIONAL PROGRAMME FOR IT IN HEALTH SERVICE

The abandoned NHS record system (NPfIT) was probably one of the most disturbing examples of a failed technology project. Costs are still vague but are estimated to be around £10 billion. The project delivered very little and suffered from numerous failures in execution. It has become an object lesson in what not to do, as there were major issues across the entire lifespan of the project. These issues include a lack of stakeholder engagement at the local level, an unrealistic timetable, testing failures, poor leadership capability, shifting aims and objectives, too much focus on price vs quality, hastily drawn up contracts, regional tender awards to a variety of suppliers, and a lack of review on the checks and balances of continuing the project. There was an attitude of 'I have started so I have to finish'.

EBORDERS

The eBorders scheme was meant to collect and analyse data on everyone travelling to and from the UK before they arrived at ports and airports. Launched in 2003, and immediately dogged with problems, it was finally terminated in 2014 after costing almost £830 million and forecasting a delivery eight years later than planned. eBorders was a hugely ambitious project involving the coordination of advanced passenger information from over 200 million journeys per year (through 600 airlines, ferry and rail carriers, and thirty government agencies). It was felt this ambition, combined with a high turnover of staff and a complex legal dispute, was the reason for the programme's failure.

WOOLWORTHS AUSTRALIA

The Australian outpost of Woolworths planned to migrate their in-house thirty-year-old legacy system into SAP. After a six year project, and going live, it became apparent each store could not generate an accurate profit and loss account. This went on for eighteen months. The root cause of the issue was a failure of the business to fully understand its own processes. The day-to-day processes were not documented and standardised, and this was compounded by senior staff leaving – taking the organisational knowledge with them.

Projects, when they go wrong, are expensive and can cause significant short- and long-term damage to organisations. The lessons learned from such projects must be baked into the projects of the future, so costly mistakes are not allowed to happen again.

Coming of age – the evolving role of technology

Having studied the legacy of technology projects, we face a fairly bleak outlook. We are in a uniquely challenging era, with many issues and problem projects hanging over us. The reputation of project delivery is poor across the industry, with the cost of failure often high and the organisational impact significant.

Despite these challenges, the digital era also presents businesses with many opportunities. For the first time, technology is driving strategic change. The capabilities new technology unlocks are largely only understood by technical teams, and these teams are spotting opportunities leaders are missing and are working out unique ways to leverage innovation. The diversity of solutions and innovative nature of the era empowers technology departments. The introduction of the CTO role in many organisations acknowledges and legitimises this. Broadly speaking – and it does differ between organisations – this role is responsible for the technologies that grow a business. By contrast, the CIO role is responsible for the technologies that support the business.

Given the changing strategic role of technology in the workplace, we just need to get better at delivery – at actually *running* projects.

Summary

The pace of change has accelerated exponentially since the Industrial Revolution, leading to the high degree of complexity characteristic of the digital era.

However, the delivery of effective change has failed to keep up with the speed of innovation we see around us, leaving organisations and individuals dissatisfied and consumers frustrated.

The technology industry has tried to resolve this conundrum by imposing frameworks on delivery through processes such as PRINCE and Agile, but these one-size-fits-all solutions require too much individual tailoring to be genuinely useful.

The time is right, therefore, to harness the innovation and creativity of technology departments to focus on delivery – on the mechanics of running projects.

CHAPTER TWO

The Digital World

What is the digital era?

The digital era began in approximately the mid 1970s. At this point, various analogue systems started to be replaced by their digital equivalents and computers gradually started to creep into mainstream use. By the mid to late 1980s the personal computer and mainframes were prevalent in many organisations, and as the pace of technology increased, the personal computer became a staple of every office desk, only to be replaced a few years later by tablets and mobile phones.

Technology projects have a chequered history and many projects have failed abysmally – the cost of these failed projects is often high, and the overarching reputation held by many senior leaders is poor. Perversely, at the same time, the digital era presents businesses with many opportunities. For the first time, technology projects are driving strategic change into organisations. The speed and innovative nature of technology present an almost infinite future of possibilities. The challenge facing many senior leaders is that the technology is only one part of the jigsaw; culture and people also make a large impact. If not managed correctly, these two factors have the power and influence to derail any solution – no matter how good the technology is.

The human factor in the digital era

Millennials

The Millennial generation is largely accepted as anyone born between 1980 and 2000. This workforce was the first to experience 'technology as standard' and typically don't have a frame of reference for the more manual practices required from previous generations. Computers and technology have always been a part of their life.

Millennials are the beginning of the connected generations. They are incredibly tech savvy and know everything there is to know about social media – because they are living it. By constantly perusing Facebook, Twitter, Instagram and LinkedIn, they have the ability to find and share information, faster and easier than ever before.

The Millennial workforce typically shares the following traits and values:

Relationships: The Millennial generation is extremely team-oriented and enjoys collaborating and building friendships with colleagues. They are more people-oriented in their style and establish close relationships. The boundaries of work and social blur easily and they tend to go on to form deep meaningful relationships within the workplace, often lasting beyond the tenure of employment.

Teamwork: Millennials see the benefits of effectiveness and efficiency over doing it alone. They prefer a sense of unity and collaboration over division and competition. As such, they tend to prefer an egalitarian leadership and are averse to hierarchies. Contrary to previous generations, Millennials were brought up in an atmosphere of equal relationships and co-decision making, and they have a community-oriented 'we can fix it together' mindset.

Immediate gratification: Millennials need to feel like what they are doing is important and that they are on the right track. This is born from the constant praise from their Baby Boomer parents. They are impatient about becoming recognised as valuable contributors and view time as a valuable resource that should not be wasted.[3, 4]

Millennials are the start of the impatient 'we want it now' generation. The product of a society that is bombarded with thousands of marketing and advertising messages a day, the average attention span since the year 2000 has reduced from 12 seconds to 8 – less than that of a goldfish.[5] On-demand services like Amazon, Netflix, Uber and Deliveroo have exacerbated this desire for immediacy. They expect instant gratification, instant answers and instant services.

High social awareness: Millennials are truly global citizens, both culturally and economically. They understand and can empathise with other cultures, are liberal in their thinking and diverse in their opinions. Many call the Millennial generation pragmatic idealists: they are aware things need to change, and of the challenges that will come with solving today's problems.

Transparent workplaces: Millennials want to feel like they have an open and honest relationship with their manager and co-workers, and that there won't be any

---

3 Gursoy et al (2008) *How Young people View their Lives, Futures, and Politics: A portrait of 'Generation Next'*, Washington, Pew Research Center www.people-press.org/report/300/a-portrait-of-generation-next Gursoy D, Maier TA, Chi CG,'Generational differences: An examination of work values and generational gaps in the hospitality workforce', International Journal of Hospitality Management, 27 (2008), 458–488.
4 www.researchgate.net/publication/44630352
5 www.time.com/3858309/attention-spans-goldfish

surprises when they join a company. They look for assurance that their opinion is valued, and tend to both give and receive a good deal of feedback. They are highly authentic, compassionate and inclusive in their nature.

Work–life balance: Millennials aren't as willing as former generations to sacrifice their personal life to advance their careers. Their motto is 'work hard, play hard', and they want to be at a company that appreciates this desire for balance. They also expect a more flexible work environment than previous generations and want an employer that supports various causes outside the workplace. They are family-oriented by nature and expect an organisation to support their family in return for their efforts at work. Flexible working is at the core of their expectations, and they expect this from the organisation they work for.

Multitasking: Millennials are multitasking experts and can juggle many responsibilities at once. This also means they can be easily distracted, finding social media and texting hard to resist. Multitasking also means they tend to attempt to accelerate their learning by doing many things at once. They want to use their time most efficiently – like by listening to self-development audio books while exercising.

Millennials: opportunities and challenges

The Millennial workforce is now entering its prime. Those born in 1980 are approaching forty, and potentially starting to make their way into senior decision-making roles within an organisation.

Project delivery needs to understand this workforce. We need to demonstrate the traits this generation expects and walk the talk. Teamwork and flat project structures need to be prevalent, and respect must be shown to time and relationships. The urgency of delivery has never been greater, but we need to appreciate the delicate balance between work and personal life.

The resources provided in project delivery must be fully connected and engaged, able to deliver to this always on workforce and compete with the distracting nature of social media and text messages. Remote working is a must. No project can be delivered without it. Relationships and responsibilities, both internal and external to projects, need to be clear and transparent, ensuring everyone knows their role and expectations. This workforce has fundamentally shifted the delivery game.

Generation Z

Millennials are not the only workforce within organisations today – Generation Z are just beginning to enter the workforce as well. This generation spans from the late 90s through to today. With their own unique experiences and values, they represent yet another challenge organisations must address.

The Generation Z workforce typically share the following traits and values:

Always connected: This generation has never known a world without the internet. The younger members of Generation Z have never known a world without smartphones. Google and YouTube have always existed. They take Wi-Fi for granted and wouldn't recognise the sound of a dial-up modem. They spend between six and nine hours a day absorbing various forms of media and 92% go online daily.[6,7] Their preferred mode of communication is digital, primarily through social media and texting, although there is a resurgence of traditional digital methods – like sending a calendar invitation to a meeting rather than a text message.

Internet experts: Always online, Generation Z find answers to questions quickly and often multitask across different screens or monitors at once. They make lightning rounds through Twitter, Instagram, Facebook, Tumblr, Tinder and Gmail. Beyond consuming information, they also create it – whether a meme, Vine, Periscope or Snapchat. They have accounts on dozens, if not hundreds, of platforms. Even putting down their

6 www.commonsensemedia.org/about-us/news/press-releases/landmark-report-us-teens-use-an-average-of-nine-hours-of-media-per-day
7 www.factsandtrends.net/2017/09/29/10-traits-of-generation-z

phone to eat is a challenge. YouTube is the new source of entertainment, documentaries and self-education – not to mention celebrity.

Entrepreneurial: Generation Z are the most entrepreneurial and disruptive workforce to date. While they may be invested in careers, they want to invent. Competitive by nature, they believe businesses can make a difference in the world and are inspired by thriving companies such as Facebook or Uber – which have literally changed life as we know it. Business Insider reported that 76% of Generation Z members aim to create jobs out of their hobbies.[8] They are better educated on planning, implementing and executing their goals and aspirations. Google is always at their fingertips. Generation Z members are also jumping on the career wagon earlier than previous generations. They are building personal brands as adolescents in preparation for pending careers. The phrase 'the early bird catches the worm' has never been more true.

The emergence of the 'gig economy' is also a feature of this generation. In an interview Mike Walsh, a futurist from Ernst and Young, speaks of the gig economy:

> 'One of the most profound changes for organizations is that most people will not be working for any particular company full time. In the United States in 2017, some 15.5 million people declared

8 www.slideshare.net/sparksandhoney/generation-z-final-june-17/2-2Meet_Generation_Z_Americans_born

themselves freelance. By 2020, some 40% of the U.S. workforce is expected to be freelancers.'

Quality over quantity: Generation Z are incredibly selective about purchases, requiring marketers to painstakingly target campaigns and demonstrate value. With abundant reviews online from people who have already purchased the product or service, Generation Z will not misuse time on a transaction they don't deem of value. In addition to quality, customisation is also a key expectation. Generation Z expects products and services to meet their exact, unique needs.

Experience over cash: Interestingly, Generation Z are less money-driven than previous generations. For them, an ideal workplace is transparent, educational and provides opportunities to advance. They want to be taken seriously, participate in senior meetings and make a difference within companies. The brand they wish to work for is open and honest, ethical and socially responsible. There is no room for a lack of authenticity; brands must tell a solid story about themselves and be highly credible.

High social awareness: Generation Z members are highly aware of our troubled planet. While most won't remember the tragedy from 9/11, they have only known a world where terrorism is the daily norm. They are passionate about change and are justice-minded, signing up for events and campaigns that previous generations wouldn't. They seek out opportunities to make an impact with a genuine intention of making a difference in the world.

'They're a hopeful generation, but realistic,' says Josh Branum, a family pastor at Faithbridge Church in Jacksonville, Florida. 'They see the world for what it is. They're not afraid, but they're going into it with their eyes wide open.'

Stressed out: Generation Z are probably one of the most stressed generations today. Technology, and in particular social media, brings with it the stresses of the 'nudge economy', where alerts sound every few minutes. This, coupled with worries about the economy, their health and the environment, make for an anxiety-ridden generation. On the plus side, they are also more likely to actually do something about these issues.

Empathetic: Generation Z are also the most accepting workforce. Norms have shifted dramatically, in many cases flipping entirely. Bullying within schools is uncool. Geeks have become popular. Prom queens are football superstars. Gay/lesbian, transgender and other sexual orientations are not nearly as stigmatised by this generation compared to others. Racially, Generation Z is the most diverse, with the vast majority having friends from a variety of ethnicities.

Generation Z: opportunities and challenges

Generation Z are entering the workforce at speed and their expectations are different even from their Millennial seniors. While generational change in the workplace is inevitable, future co-workers and employees will test the limits of existing management models, as well as challenge traditional beliefs about company culture.

Generation Z demand the drive for continuous disruption and reinvention. They expect organisations to be highly connected, while simultaneously having a deep social backbone that influences the world in which we reside. The 'gig economy' requires organisations to communicate better with new hires and temporary resources alike, and for managers to bring teams together far more effectively. This generation also needs to be let loose entrepreneurially within organisations. They must be given the freedom to create under the auspice of guidance and mentoring. They want to make a difference.

Organisations will need to communicate differently, on different platforms, and be sensitive to individual expectations and career aspirations. They must also juggle priorities in terms of work–life balance, social interactions and quality over profit.

Organisations that facilitate the success of the Millennial and Generation Z workforces based on the characteristics above will undoubtedly flourish. Culturally, organisations must adapt to the new workforce. They must change the very nature of how they perceive work: their survival depends on it.

Problems and issues created by the digital era

There are a number of problems and issues we must resolve in the digital era to be successful:

- **The speed of change:** The speed of technological change continues to accelerate. Companies first to the market with good products, facilitated by technical solutions, will dominate. Companies have the ability to appear out of nowhere and elevate themselves to an almost untouchable level. Both Uber and Airbnb appeared almost overnight and swiftly dominated their sector. These organisations are capitalising on technology and are leveraging it to competitive advantage. Legacy processes need to improve and take advantage of their digital equivalent – often being enhanced by technology, intelligence, machine learning and/or artificial intelligence.

- **Cost**: Traditional organisations are more cost-conscious than ever. They understand the need to invest but are risk-averse and cautious. The failures of the past make traditional organisations concerned about technology projects. Interestingly, those start-ups leveraging technology for competitive advantage are the opposite – willing to spend to get ahead or for a foothold in the market. Both want to see a healthy ROI.

- **Complexity:** Technological advancements, combined with layering different technologies, present new opportunities. Organisations need to react swiftly to spot these emerging opportunities; they may lead to new products, or even new markets or segments.

- **Expectations:** The expectations of organisations seem to escalate with each passing day. New technology, harnessed well by leading organisations, sets a precedence for others. They, in turn, expect more from their own projects, without understanding the difference in organisational design, history, culture and readiness.

- **Millennials:** The Millennial workforce is now beginning to enter senior management. They have different ideas and ethos compared to the Baby Boomers before them. They grew up in a different world and have different cultural expectations and principles. The books on leadership are beginning to be rewritten by this generation.

- **Generation Z**: This generation are just beginning to enter the workforce, but are presenting yet another shift in priorities compared to their Millennial leaders.

They require a second look at how we manage our teams and run our businesses using this new group of talented individuals.

Consequences for business

Broadly speaking, businesses have two choices: adapt or die. We are already seeing the demise of those who focus too heavily on their heritage and history rather than the demands of the future – often highly established brands. They are looking backward rather than forward. While heritage and history are important to some, the new, more informed customer wants a product and brand that fits into the current world. One that does what it does well, rewards and treats colleagues as expected, and that gives and is socially responsible. The market will naturally favour those that embrace change towards this ethos and they, in turn, will rise to the top of their industry.

> It is not the most intellectual of the species that survives; it is not the strongest that survives; but the species that survives is the one that is able to adapt to and to adjust best to the changing environment in which it finds itself... so says Charles Darwin in his *Origin of Species*.
>
> —Professor Leon C. Megginson, Louisiana State University

Project leadership in the digital era

What does this mean for our project delivery and orchestration? Quite simply – a great deal. We need to imagine our projects as complete business ecosystems. This ecosystem requires solutions and considerations for each of the points raised regarding the digital era, as well as focus on the outcome and output of the project in order for everyone to move in the same direction. As Patrick Lencioni says: 'If you could get all the people in an organization rowing in the same direction, you could dominate any industry, in any market, against any competition, at any time.'

The modern project leader needs to understand both the micro and macro level in which they are operating. The tools and techniques provided by industry focus on the tactical or the micro. Fundamentally, it's how to deliver the project – the very mechanics of project management. The macro focuses on the traits of the era they are operating within: how to make the most of the digital era by managing the pitfalls and taking advantage of the traits we are now seeing. If the project leader can master both the micro and macro, they are capable of running truly remarkable projects.

PROMPT – accelerated delivery in the digital era

The digital era, and all of the associated problems, pitfalls, and benefits, present a headache to the modern business. Projects must have well-orchestrated governance and control, they must deliver on highly aggressive timelines, and they must offer a good return on investment. In researching this book, and speaking to hundreds of companies about their issues and desires, a common theme emerged:

> If we can run a highly accelerated project under controlled conditions, we will deliver things quickly and safely.

The natural result is a reduction in cost, as typically the largest cost to projects is people.

The PROMPT accelerator was created to address this principle. We took our research and analysed what it takes to run an accelerated project. We then overlaid this with the traits of the digital era, noticing an almost direct correlation. We closed any gaps with numerous interviews with business leaders and grouped the information we gathered into 'common sense' categories. What resulted are principles and techniques that overlay some of the more traditional transactional (or micro) delivery models. This is not meant to replace them, but rather to improve and supplement while providing an additional structure around getting the job done. Quickly.

The PROMPT acronym stands for:

Preparation: How do we prepare ourselves, both as an organisation and as a project, to run an accelerated project?

Reorganisation: What do we need to improve, remove and change in order to run an accelerated project?

Outcomes: How do we shift the project to one that is outcome oriented, rather than output oriented?

Mindset: How do we construct and promote a mindset to deliver accelerated projects and winning teams?

People: What do we need from our teams to deliver accelerated projects?

Tempo: Once we have set the required pace of an accelerated project, how do we maintain that momentum throughout the duration of the project?

The rest of this book explores these topics in detail to uncover the secrets of running an accelerated project in the digital era.

Summary

The digital era is characterised not only by the technological capacity of organisations and individuals, but also by the context in which they are operating.

Consumers, clients and workforces present a very different profile, in terms of their aspirations, purchasing habits and leisure pursuits, than they did even ten years ago. If organisations are to thrive in the digital era, they will need a good understanding both of who their 'audience' is, and of their own people, to track what is essentially a moving target.

The PROMPT accelerator is the approach that will enable businesses to keep up, as it enables controlled delivery of projects to tight timescales, resulting in reduced costs and responsive solutions.

CHAPTER THREE

Preparation

Preparation is everything. There is not a single area in life where preparation does not make a significant difference. A professional footballer would never play a match without warming up, a lecturer wouldn't speak without prepared content, an entrepreneur wouldn't launch a business without first testing an idea. Projects are no different.

Believing that we can start a project and make it up as we go is beyond ridiculous. Projects are difficult and will only be made more difficult without defined controls on the temporary environment. Planning and preparing will pay massive dividends and be a huge contributor to the success of your project. Before you even think about a business case and benefits, start preparing for the project's execution and requirements.

Lean into the preparation phase – and make sure you specifically carve out the time to do so. Once senior leaders are involved, expectations will be set and delivery expected. At that point, like it or not, you are in delivery mode. The time for thinking conceptually has passed – and you should know exactly what you're doing and how it will be done. Even if it's only theoretical, you should know the approach you're going to take. Making in-the-moment decisions or getting caught on the back foot is a slippery slope to failure. Like Bear Grylls and the Boy Scouts – be prepared.

Setting compelling goals and objectives – the why

The first step in preparing for a project is to ensure you are crystal clear on *why* it's happening in the first place. What's the goal you are set to achieve? What end state are you aiming for? Determining what your why is, and communicating it clearly, is vital to any project. It motivates the team, allows prioritising among other projects, and helps provide clarity on decisions. It will give all team members and stakeholders a sense of purpose and belonging.

Do not confuse your why statement with a list of objectives. Your objectives will come about because of the why. Typically, the why is broad, empowering, visionary and compelling. There are some great books and techniques for defining your why – one of the most powerful being Simon Sinek's TED talk 'Start with Why'. Granted, he is thinking more in terms of starting a business, but a project is no different. Projects must be viewed as business ecosystems. Get the why – or the reason for being – right, and the associated objectives should flow steadily and easily. Jumping straight into objective setting and metrics often leads to quite narrow thinking and doesn't allow for creativity.

Consider the following: a global company is reviewing how it provides back office support to the operational or revenue-generating parts of the business. If we take an objective-led approach, we might see entries like 'implement a new forecasting tool for the next financial year' or 'train all management in the importance of forecasting ready for the new financial year' appearing in the business case. These will have already shaped the outcome and hence constrained creative thinking. Using the same scenario, an appropriate why statement might be, 'We are going to create a supporting finance infrastructure that drives value into the heart of all operations, where there are no surprises, and finance is accessible and meaningful to all.' The latter is much more appealing to end users and gives scope for broader creative thinking on how to solve the same problem.

Once you have your why, begin to expand it into specific, measurable and time-bound objectives. The objectives will occur as a result of the why. Some may be obvious, for example, 'reduce the time taken to close month-end', but others might be slightly more unusual, like, 'improve staff retention in finance (better process = happier teams = less turnover)'. The objectives are ultimately what the project will be judged against. The most successful projects begin thinking about realising these benefits early in the delivery cycle, never losing sight of them during decision making. Third-party auditors, if involved in the project, will certainly refer back to the initial objectives. Think of them as the scorecard for the project. This is even more critical in the case of publicly listed companies, as shareholders will expect a return on the project investment. Objective or benefit mapping needs to feature early in the life cycle of the project to prepare the team and set expectations.

Once defined, make sure everyone knows and understands what the why of the project is. The project will live and breathe this mantra so everyone should, at an absolute minimum, know what the why is. The objectives may be required for measurable detail, but the why drives the team. For the duration of the project, that why is the reason for being.

Do not do anything, or even think about starting something, until you are crystal clear on your why and its objectives. Make sure they are understood and agreed to by everyone. This is the whole purpose of your journey for the next few months.

TOP TIP: Objective setting is all about outcomes and results, not about activities or tasks. For example 'Hold two membership drives per year' is an activity-based objective and therefore a poor one. The appropriate objective would be 'Increase membership by 30% over the next twelve months'. A good objective should be achievable by a number of routes, encouraging creativity and free-flowing ideas.

AN EFFECTIVE WHY STATEMENT

A large client was running a substantial global SAP roll-out (predominantly finance based). The why statement they settled on was 'to remove barriers, allow creativity and innovation, empower all finance colleagues and allow them to fulfil their potential, supported by best in class global technology solutions'. This provided an initial reason for being and acted as a guide for the remainder of the project. It forced the team to consider approval levels (empower), steered them away from their existing processes (innovate), and pushed them towards clever and unique solutions (creativity) capable of operating processes globally (global technical solution). The result was one of the best and most elegant finance solutions ever seen.

Definition of success

Now that we have our why and our objectives, we should have a clear representation of what the project's success looks like. We should understand the future state organisation we are aiming towards and the operating model it will transact through. Some of this may be theoretical at this stage, but the principles or foundations should be visible. The why, together with the objectives, form the definition of success. In an Agile framework we would call it the Definition of Done. There may be occasions when the objectives are sensitive in their nature: they may, for example, include changes to headcount. If this is the case, the best approach is to be bold and transparent regarding this from the beginning.

This is a totally different way of thinking, but transparency and straight-talking, no matter how difficult, will always be better received than a hidden agenda. It takes strong leadership, but everyone will appreciate clear messaging from the outset. Remember, you are operating with the best of strategic intentions for the company. You are safeguarding it for future generations and workers. The leadership and project team will need to be aware that the messaging and communications surrounding the project, unless they are sensitively managed, have the potential to disrupt workers. The more communication the better, even when there is no news. Keep control of the 'water cooler conversations' and use your business networks to test the feeling within the business teams. Encourage everyone to look forward to a future that looks different, and build commitment to that vision. Those who are fully committed to the organisation will prove their worth in the coming months.

Communicate, communicate, communicate

Once the why, objectives and definition of success are settled, communication is key. These components should be understood by the entire organisation, both for those individuals within the project, and those that may come into contact with it. You may even consider sharing it with relevant third parties. These statements will drive all decision making from this point onward. Make posters or create desktop wallpapers and screen-savers. Ensure they feature in corporate newsletters and regularly appear on bulletin boards or corporate internal blogs.

Once the first round of communications is complete, schedule a regular refresh of these messages. They are the foundations for organisational change and, considering the project will cause some short-term change pain for certain groups of people, the why message needs to be constantly reiterated.

TOP TIP: One of the best approaches to continuous change messaging is to set up a blog series interviewing key (typically senior) colleagues and asking them to explain the project in terms of the impact it will have on their departments or organisational units. This provides fresh content, can be created at the outset, and allows the audience to understand the project from a multi-dimensional approach. It will be important for the organisation to understand that the project is not just a sales project, for example. If you are feeling brave, spin this into a video series!

Set up the right strategy and tooling

The strategy and tooling phase is where you will really have to get your hands dirty, as now you are designing how your project will be executed.

What is the point of a strategy document?

Strategy documents provide a reference point for the organisation on how the project will operate. They are typically agreed by the stakeholder team and control the journey the project will move through. They are excellent for onboarding new team members and provide a good foundation for 'how things get done'. They also document any decisions made around policy and process for the subject area – for example, how and what data you will be importing into the new system, and what the rules are for each data object.

Strategy documents

At a *minimum,* you will need the following strategy documents to execute the project:

- **Design and blueprinting:** This will govern how you are going to run the design or blueprinting phase. How will you capture requirements? How will they be documented? Will you run workshops or analyse business processes in situ? The suppliers (either internal or external) will have a list of things they need to know to deliver and/or build the system. The purpose of this is to document the approach taken for gathering all of this information. It's important to lay this out from the outset (before starting the work), as it sets expectations and allows resources to be released from their usual responsibilities. It also starts to control the level of decision making required by the project. For accelerated projects, we must push the decision making down the hierarchy and let the specialists doing the work make the right decision for the project.

- **Adoption vs adaption:** This is often part of the design and blueprinting document, but we like to pull it out separately as it's critical to accelerated programmes. If you are buying an off-the-shelf product or a product that can be highly configured (such as an ERP), the closer you stay to the basic product the faster the implementation will be.

- **Design authority:** This is another element sometimes included within the design and blueprinting strategy, but again, it's important enough to stand on its own. The design authority is the team of people who sign off on the concepts of the build and any future changes to it moving forward. The process this occurs through is documented within this strategy.

- **Landscape and client strategy:** If you are managing a software development project or an ERP project, you will need several instances of the 'build' to complete various project tasks. Development and unit testing will typically occur in the same instance, but the system or user testing (where the build needs to be more stable) usually happens in a clean environment. The challenge is how to keep the two in sync with code change control, etc. This strategy governs the landscape, explaining what happens in each of the builds and how change is controlled (ideally through an automated tool). Running a highly accelerated project requires more systems to cope with agility as well as tooling to move data around the different assets. The process for this and its requirements are normally documented here.

- **Reporting (end user):** This may fall under one of the design documents, but it's important to start thinking about the outputs of the system before the project begins, especially as too many projects have failed to deliver reporting correctly. This document should lay out the principles for reporting: where will it be done? What technology will be used? At what level of detail? How will it be updated?

- **Testing:** Understanding how testing will be carried out in the preparation stage is crucial. How many phases will you have? How will you determine test scenarios and scripts? How will you log progress? Where will you store evidence? What are the entry and exit criteria for each test event? Will you be doing any non-functional testing? Accelerated projects need to work through this phase of the project quickly – they need to fail fast. Testing is one of the biggest drains on time, and arguably one of the most critical. This phase needs to be well-orchestrated and run as smoothly as possible.

- **Data migration:** Another crucial component is data migration – especially as this can be quite a contentious document. This is the overall strategy for how you are going to handle data. It should detail what data objects you will be migrating into the new system, and to what volume. Will you be migrating monthly balances or full transactional data? How many years of customer data will you bring across? What about their sales history? What are the legal requirements? This document will also cover the number of iterations of data loading the project will go through (they never work the first time!) and cover how the data will be reconciled from the source to the target system – and who will do the work. Lastly, and equally importantly, it will explain the data cleansing process. The legacy data will likely need to be cleansed, updated or enriched before it is suitable for inserting into the target system.

- **Training:** This document explains how the user base will be trained, how the training material will be delivered, and any systems or technology required for the education phase of the project.

In addition to the above, the following documents will be required to manage and govern the project:

- **Documentation approach:** Just because you are implementing at speed doesn't mean you can slack on your documentation. This strategy explains what is required, the standards and where it will be stored.

- **Planning:** This is the overarching guide to how the programme will be planned and who will do what. It explains the various levels of plans within the programme.

- **Governance:** This will explain the governance process, who is involved, and what the roles and responsibilities are. It should also frame the decision-making capabilities. To accelerate projects everyone must lean in on decision making and push decision making down the project hierarchy. If the project is complex, this document may also explain how the team will be structured and how reporting lines operate.

- **Project reporting:** This will explain how, and to what frequency, project reporting will occur. It also explains the level of detail required in each report and who the target audience is.

- **Change request strategy:** The project, especially in the digital era, is guaranteed to be subject to change. The approach to change management must be fast and robust. Everyone in the project, including the senior team, must agree on the approach. This strategy should cover the process, approval steps, impact analysis and cost approach.

- **Risk and issue management:** This strategy details how the project will manage risk and issues. It outlines the mechanism for management, the frequency of assessment and the prioritisation process.

- **Benefit realisation:** Any new system, technology or otherwise, doesn't realise benefits on its own. There is often additional work involved in reorganising the organisation to leverage the anticipated benefits. Change management is key, and benefits from the initial objective statement should be managed through to delivery. This strategy explains how this will occur and who will do what.

- **Communication:** This is the strategy document for how the project is planning to communicate both internally to the project team, externally to the wider organisation, and to suppliers and customers.

It typically contains methods and channels as well as the tone of the communication.

- **Stakeholder management:** This explains how you will manage your stakeholders. It categorises them and explains, via a responsibility model, how they will be involved in the project. This helps to drive various aspects of the project and ensures everyone is clear on the levels of engagement with those involved.

Tactical documents

If the strategy documents are the *how* of the project, the tactical documents are their corresponding deliverables. These will be used daily and will help manage the project to completion. They are typically varied depending on the solution being delivered and the delivery methodology, but may include:

- Test log and system
- Testing entry and exit criteria
- Test exit report
- Data migration templates
- Design and Blueprint template
- Adaption template
- Transport request form
- Functional specification template
- Technical specification template
- Baseline plans
- Highlight Report template
- Readiness assessment

- Change request form
- Training templates
- Standard Operating Procedure template
- MS Office templates
- Meeting agenda and minutes templates
- Roles and security matrix
- RAID log
- Report designer templates
- Deliverables by week and month
- Benefit tracker
- Communications template
- Newsletter

At this point you're likely panicking over the volume of work required to control and execute a project, let alone one that you're doing at speed. However, remember two things:

1. The project may be one of the biggest the business has implemented and is perhaps an evolution or even revolution to what they do at present. It needs safety, stability, predictability and control.

2. Speed does not equal a lack of control. Running an accelerated project is not a get out of jail card for sloppy projects. There is a lot of misconception that accelerated projects (using methods such as Agile or DevOps) are chaotic, undocumented and out of control. It could not be further from the truth. The best practices still apply. An accelerated project just means you do more in a shorter timeframe.

It might be that you have corporate or departmental policies for some of these strategies. Rather than blindly adopt them, review and challenge them. Can they be improved? What is the process for each strategy? Is it required? What is the minimum requirement to maintain control?

With each of the documents you create or amend, think back to the why statement and remember your project is about acceleration. Needless processes burn the one thing you can't control: time. Ensure your processes are as lean as possible while still maintaining the level of control you require. Think about the minimum viable product concept. What gives the control your organisation needs, at the leanest possible level, thus preventing a loss of velocity in the delivery?

Consider the same key points with the tactical documents and tooling. What can you put in place to accelerate your delivery? Will it let you start fast and stay fast? Each project will be different, but consider tooling vs the alternative approach. Work out the associated ROI of each and judge the route accordingly.

Some of the best projects we've seen were so successful because they had the strategies and tactical documentation in place before they even started the project delivery journey. Everyone was on board with the processes and the roadmap of the project before the start gun went off. One client, in the public sector, was so organised in their preparation that they instructed their delivery partner on what they wanted to occur during the project. This made it simple for everyone. The result was a project delivered early and under budget – and with an enhanced scope.

Tooling

Decide what tools you will need to operate your project at a high velocity. There is lots of software available to help run accelerated projects and many are a SAAS model (software as a service) in the cloud and can be activated for the lifecycle of the project only. The more automation you can build into your project the better. Software packages such as Slack, Basecamp, Trello, Microsoft Teams or even Microsoft SharePoint help projects run better. If you're running an ERP project, work out where the bottlenecks will be and look at what software could help. Assess the complexity of the project, pick your software and commit to it fully.

If you do go down the software route (and you should), bear in mind you will need to include this cost in the funding for the project. You will also need to consider the time required for adopting the new software: a short familiarisation period will be required before your teams understand the software and are comfortable using it – then it will become an asset to the project.

Roles and responsibilities

Projects are temporary organisations, or a mini-business, if you will. They have a why, objectives, goals and processes to deliver, and a customer who receives the output.

Continuing the analogy, you wouldn't bring someone new into your business without onboarding them or explaining and defining their role carefully. If you didn't, how would they know what their role was? How would they know how they fit into the wider organisation? Or what the goals of the organisation are? The same is exactly true with a project.

Every person the project interacts with needs a role. They need to understand exactly what is expected of them, what they are accountable and responsible for, what they will do on a daily basis, what decision-making powers they have, and absolutely critically, *what they do not do!* The latter stops people interfering and slowing processes down. It stops people taking over. The principles around accelerated projects are that everyone knows that they are doing and what everyone else is doing. People taking an interest in something that is not their area of responsibility or accountability leads to excess meetings, too much discussion and procrastination – and not enough delivery.

Decision making

As part of the roles and responsibilities, you must be crystal clear on the levels of empowerment and decision making you are giving the team. This is brutally simple: the more decision-making power you push to those delivering the project, the faster the project will progress. This is scary stuff at first, especially for senior leaders, but frame it like this:

Members of the project team are all working towards a common why, and the why is linked to objectives or a success statement. These form the ultimate target against which everyone in the project is being measured. If required, this can be reinforced through the creation of a compelling reward structure. Everybody wants the project and the project team to succeed. Human nature pushes us all to win and succeed, not lose and fail. The project team and the accountability of the team keep individual decisions in check: the more you give, the more you get.

Training

When you organise your colleagues into a project, give them new roles and responsibilities along with enhanced decision-making capabilities. As per the previous analogy, this is no different from joining a new company. During the onboarding process, you need to coach and train these individuals so that they are empowered and sufficiently skilled to succeed in their new roles. From the project perspective, the more skilled they are, the better they will perform – and the project will run faster. From a health perspective, the more comfortable they are, the less stress they will feel – and the more productive they will be. Win–win. Make sure you are looking after your team.

LOOK AFTER YOUR TEAM

While working with one client, they decided to take the entire team to stay in a hotel for a week. While away from the distractions of their usual day job, they taught the project team everything they needed to know about running the project, demonstrated the solution to them, and gave them a world-class onboarding experience. The result was a highly motivated and highly knowledgeable project team, who really understood the journey they were about to embark on.

Environment dictates performance

There is a lot of research on the various environmental factors that will impact performance. Entrepreneurs and business leaders perform better in ecosystems of like-minded people. Sportsmen and women play at their best when they are pushed by a coach or highly skilled team mate/opponent. It follows that surrounding a project team with other high performers drives an increase in performance, no matter what the task.

This also extends into the physical location. If you give an athlete access to the best facilities they will improve and perform in proportion. The teams in some of the world's biggest and most successful companies do not work in tired or old-school office environments. The same approach should be considered for a project team. You are expecting the team to perform at the highest level and you want them to do the best work of their careers for you. It stands to reason, therefore, that to facilitate this you would give them an environment they need to be inspired and to succeed.

Clearly we live in a global era: co-locating may not be an option, so look at other tooling to facilitate tight collaboration and leverage this. The goal should be to create a close knit community of highly skilled experts who have everything at their fingertips to enable them to succeed. The acid test is the level of desire you create. You need everyone to *want* to be on your team.

Planning

Planning is crucial to any project, but even more so to one that is highly accelerated. Any deviation needs to be absorbed quickly and issues resolved urgently to keep the project on track. Any obstacles need to be identified as early as possible and mitigated swiftly. Working at pace leaves little time to reflect on the bigger picture, and sometimes a single day's delay can be critical.

Plans should be baselined before work starts and should be readily available to the team. Learn to utilise the skills of the project management office (PMO). A PMO is typically the governance, reporting and administrative hub for the project. They own the plans and need to live and breathe them. Get into a habit of continually reviewing plans. Some projects require a daily check-in with the PMO, others weekly. Fit the requirement to the project and monitor, monitor, monitor. Consulting the plan regularly will allow you to identify early when things aren't on track. The earlier you spot an issue, the longer you have to resolve it.

Plans also need a dose of realism. An overly optimistic plan will not deliver and a pessimistic one will not move any further than the business case. It's a fallacy to believe a project will operate faster just because the plan says so. Projects are accelerated if and only if the many aspects of the project are structured to be accelerated. They are not accelerated by doing more of the same but with an optimistic and unrealistic plan.

As Einstein said, 'Insanity is doing the same thing over and over again but expecting a different result.' (Interestingly, this is commonly misquoted and there is much debate about who said it first, but you get the idea.)

Ambition vs realism

Depending on the component of a project, you can set an aggressive timescale and achieve it – contrary to what the project team may initially think. This is where the skill and experience of the management team are vital. The leadership team needs to know when (and what) to stretch for and when to consolidate. Push too hard for too long and the team will collapse under the pressure. Sit and consolidate for too long and you may deliver to plan – but you won't delight a customer by delivering early. A good project manager likes to create slack in their planning by pushing then holding, pushing then holding. This builds additional contingency or early delivery. Both of which can be leveraged throughout the project.

Summary

Just like a house, accelerated projects need foundations. The more solid the foundation, the easier the house is to build and the longer it will last.

One of the key factors to consider here is that we still haven't even started the project. While there may be a PMO cost in facilitating the above, we still are to fire the starting pistol. There certainly shouldn't be any supplier costs at this point. Get your house in order first.

Preparation checklist

- ✔ My project has a strong why statement.
- ✔ It has objectives and success criteria that are specific, measurable and time-bound.
- ✔ We have defined all our strategy documents required to run the project.
- ✔ We understand the tooling we will use to run the project.
- ✔ We have researched and chosen any software needed to help run the project and have factored it into budgets and timescales.
- ✔ People understand their roles and responsibilities and have been trained appropriately.
- ✔ Decision making has been pushed down the project hierarchy.
- ✔ Plans are in turn realistic or ambitious where required.

CHAPTER FOUR

Reorganise

It is highly likely that there will be a number of incumbent processes, methodologies, or 'ways of doing things' within your business. These will have evolved throughout previous projects and adjusted and grown organically to suit the needs of the organisation as it moved through various growth cycles. The challenge in the reorganise phase is to look at those processes, and like a champion gardener, prune back the branches that have overgrown. The gardener knows this pruning is in the best interest of the garden overall. The same is true for the organisation: outdated or tedious processes ultimately eat time.

Accelerated projects work best when the processes involved are flexible and simple: but this is not an excuse to cut elements from a project. There are far too many project leaders citing a lack of documentation and control on Agile, and iterative or accelerated delivery methodologies. The choice of methodology is irrelevant – there are NO excuses for cutting corners. The correct execution of the right lean processes and project leadership is what matters.

What is simplicity?

There are many of definitions of simplicity. The one we prefer, and the one that most closely aligns with our actions in this phase, is credited to John Maeda – a leading force in design-based simplicity. He is an executive, designer, and technologist working and exploring the area where, in his words, business, design and technology merge. In his book from 2010, *The Laws of Simplicity,* Maeda refers to ten laws and three keys.

Maeda's ten laws:

1. Reduce – the simplest way to achieve simplicity is through thoughtful reduction

2. Organise – organisation makes a system of many appear few

3. Time – savings in time feel like simplicity

4. Learn – knowledge makes everything feel simpler

5. Differences – simplicity and complexity need one other

6. Context – what lies in the periphery of simplicity is not peripheral

7. Emotion – more emotions are better than less

8. Trust – we have higher levels of trust in things that are simple

9. Failure – some things can never be made simple

10. The One – simplicity is about subtracting the obvious and adding the meaningful

Maeda's three keys:

1. Away – more appears like less by moving it further away

2. Open – openness simplifies complexity

3. Power – use less, gain more

Exploring and translating these concepts in our process-driven world encourages the analysis and challenge of our existing processes and provides productive ideas on improvement. We need to explore the meaningful concept of any process and challenge the value it adds. Often the process can be dissected into inefficient components. Our goal should be the removal of these, leaving only the areas that add value with the people who add value.

Process-driven simplicity

Similar to Maeda, the Japanese – through their manufacturing industry – are some of the best in the world at driving simplicity into business and processes. There are many models, all falling under the bracket of 'lean' with roots in the Far East – Kaizen, Kanban, Six Sigma, Total Quality Management, 5S, Toyota Production System, etc. There are numerous highly detailed books on each, but all broadly focus on standardisation and efficiency, and the removal of three types of waste: *muri* (things that are too difficult), *muda* (things that are wasteful), and *mura* (things that are irregular).

We are going to look at three of these models in a little more detail – Six Sigma, Kaizen and 5 Whys. These are some of the most well-known, and the concepts will help to spark ideas on how you can improve your internal processes to speed up projects. Both Six Sigma and Kaizen are Japanese influenced, but they are two different programmes. Kaizen is predominantly focused on continuous improvement (eliminating waste and ensuring efficiency), while Six Sigma's focus is on eliminating defects and reducing variability. The 5 Whys actually features within Six Sigma, but is worth calling out separately as it's a great technique for drilling into root cause analysis.

Remember: the more you can increase speed and decrease waste and inefficiencies, the faster your project delivery capability will be. Use the techniques below to analyse your existing organisational processes for delivery, and identify what you can improve on.

Kaizen

Kaizen is based on the philosophical belief that everything can be improved upon. Nothing is ever seen as a status quo; there are constant and continuous efforts to improve, which result in small, often imperceptible, changes over time. These incremental changes add up to substantial changes over the longer term, without having to go through radical innovation. This can be a much gentler and employee-friendly way to institute changes that must occur as a business grows and adapts to its changing environment. If a work environment practices kaizen, continuous improvement is the responsibility of every worker.

Much of the focus in kaizen is on reducing 'waste' through several forms:

- **Movement:** moving materials around before further value can be added to them.
- **Time:** spent waiting (no value is being added during this time).
- **Defects:** which require re-work or have to be thrown away.
- **Overprocessing**: doing more to the product than is necessary to give the customer maximum value for money.
- **Variations:** producing bespoke solutions where a standard one would work just as well.

Avoid the pitfalls above during your project processes, both while preparing for the project and executing it. Look for ways to improve processes and remove waste on a daily basis. Empower your team to make changes for the better – encourage them through project improvement sessions. Even at the micro level, this soon scales to be significant.

Six Sigma

Six Sigma is a set of techniques and tools for process improvement, created by Motorola in 1986. It seeks to improve the quality of the output of a process by identifying and removing the causes of defects, and minimising variability in a process. It typically uses a set of empirical and statistical methods and utilises a special hierarchy of people within the organisation who are experts in the methodology.

The term Six Sigma originated from terminology associated with statistical modelling. The maturity of a process can be described by a sigma rating indicating its yield or percentage of defect-free products it creates. A Six Sigma process is one in which 99.99966% of all opportunities to produce an output are free of defects. That's three out of every 1,000,000 opportunities – statistics anyone would aspire to.

Six Sigma projects follow one of two project methodologies: these methodologies, comprising of five phases each, bear the acronyms DMAIC (Define, Measure, Analyse, Improve, Control) and DMADV (Define, Measure, Analyse, Design, Verify).

DMAIC is used for projects aimed at improving an existing business process, while DMADV is used for projects aimed at creating a new product or process design. Six Sigma is highly statistical and evidence-based – everything ties back to a measurement taken at the beginning of the process. This appeals to and persuades many sceptics, especially those of an analytical mindset.

The '5 Whys'

Put simply, the 5 Whys is an iterative interrogative technique used to explore the cause and effect relationships underlying a particular problem. The primary goal of the technique is to determine the root cause of a defect or problem by repeating the question 'why?'. Each answer forms the basis of the next question. The technique is simple but highly effective. It cuts through politics and gets to the root of requirements and processes quickly – enabling analysis of the situation and ideation towards achieving a similar outcome, but in a more effective and accelerated manner.

USING THE 5 WHYS TO IDENTIFY THE PROBLEM

One client ran this concept against their change management approval process. The results are surprisingly obvious, but the 5 Whys drives clarity and demonstrates the importance of the solution derived, and its impact.

Problem: Our change management takes over a week to approve, and we don't start building until everything is approved.

Why?: Because the workflows in the system are not completed in a timely manner.

Why?: The workflow owners are not happy with the level of documentation regarding the change.

Why?: There is missing documentation on certain prerequisite items.

Why?: Users have not uploaded the documentation.

Why?: We haven't told everyone what is formally required.

Why?: We thought everyone knew already.

Solution: Document and communicate the process and requirements for change management documentation, providing templates of each document to ensure consistency.

Regardless of your choice of methodology, the principles around lean or continual improvement remain the same: we must analyse our processes to reduce the time taken to execute them. A project held up by process has already stalled. Encourage your team to drive out the waste and optimise your project environment.

TOP TIP: If you embark on a process of continual improvement, ensure you follow through and execute the suggestions. If you ignore a good idea, it has the opposite effect and demotivates team members. Remember the golden rule: your team knows best!

Embedding simplicity through your project

Focus on embedding simplicity into the lifeblood of your project – and shout about savings and improvements you are making to processes! Typically processes frustrate colleagues, especially if they are trying to run faster than the process allows. Showing a culture of improvement and listening is crucial. It motivates colleagues, speeds up processes and demonstrates that the leadership is listening and doing all it can to remove bottlenecks.

It is worth considering running a regular section of a project update meeting on improvements – listen to your team's suggestions and follow through on any promised improvements.

Areas ripe for process improvements

There are several areas that come up regularly in most projects that can benefit from improved process:

Build: change control, costing, functional and technical requirement capture.

Testing: test cycles, defect process, retest process, test reporting, testing coverage, test automation.

Deployment: movement of change, deployment live.

Governance: reporting, costing, change control, planning.

Communication: internal and external project communications, meeting frequency.

Training: training delivery, training collateral.

This list is by no means exhaustive. Speak with leads for each area to explore their views on what could be improved – they are the experts.

> **TOP TIP: Look at the projects your organisation has executed previously and work out the pain points. History teaches us much about our weaknesses.**

Controlling simplicity without cutting corners

By removing waste and inefficiencies, you will decrease the time required to execute and increase the overall value of the process. The goal in our accelerated world is a process that is meaningful and fast – resulting in faster projects. This is not about removing something from the process to save time. A common trend, and indeed a big mistake, is the removal of a component of the process to save time and ignoring the value it adds. There are many examples of accelerated projects failing to document requirements of the build, all in the pursuit of time. These things inevitably come back to haunt projects and often end up burning more time than they save. Accelerated projects are not an excuse to ignore key value-added processes.

MORE HASTE, LESS SPEED

One client with a project in distress went through a turnaround phase with us. After analysing the processes the project was operating with, they were found to be far too complex and lengthy. This was caused by senior management adopting a command and control approach to nurse the broken project. The time pressure these processes caused forced individuals to cut corners with standards and documentation. The project went live, but the number of support tickets was high and performance of the system dire. In driving through to the root cause analysis (using the 5 Whys strategy), it was discovered that none of the coding standards had been adhered to and no documentation existed – all in the pursuit of rushing and cutting corners. This was, ironically enough, driven by the excessive processes that had been put in place to control the project. The lesson? Processes need to add value and shouldn't impinge on timelines. Rushing and cutting corners helps nobody.

Challenge the need for ad-hoc in-project change

This is probably one of the most important sections in this book, and if managed correctly will keep the build simple and cut both time and money from any project.

The three rules

There are three rules all projects should adhere to:

1. If developing from scratch, agree on a scope (or a series of stories), and stick to it. Deviation or changes, at the fundamental level, hurt project delivery. If you are working in an Agile or DevOps environment, having no scope creates a never-ending project. You must define the project and create an end state.

2. If implementing a COTS (commercial off-the-shelf) product, scope it fully before purchasing it, then implement and use it as it was intended. Do not bend the product to fit your business needs. Find another product or an add-on.

3. If implementing an ERP (including cloud-based), accept the product as it is delivered. ERP projects suffer severely from change syndrome. The more change is made to the core product, the costlier it is – both in terms of the project and the total cost of ownership. It also becomes more challenging (due

to multiple development streams) and more time is wasted implementing and supporting it. Just because you *can* change it, doesn't mean you *should*.

While the first two points are fairly self-explanatory, point number three is the cause of many sleepless nights for CFO's. Indeed, there are many unnecessary horror stories of failed ERP projects that couldn't control their scope and were ultimately deemed not fit for purpose. We need to understand why this is the case and protect against this project killer.

ERP – prevent a potential nightmare

An ERP is a configurable item. It can operate in many different ways to provide the answers to many different process problems. Additionally, there are numerous exit points written into ERP software which allows for custom code. This, in theory, means we can make the ERP solution do exactly what we want it to. This probably rates as one of ERP's biggest strengths. However, and it is a big however, this strength has fast become one of its biggest weaknesses. With infinite flexibility, organisations struggle to leave their existing business practices in their old system and adopt new processes in the new one. Instead, they re-engineer them inside their new ERP – stifling innovation.

Let's pause for a moment to understand how ERP evolved. Over the last thirty or so years, various companies built big ERP systems capable of running entire businesses. These were adopted by many businesses, who in turn contributed and innovated additional ideas. Those considered good ideas were built into the core product for the next release of the software. The cycle of adoption and contribution continues, and now companies like SAP and Oracle have a product that is the result of 100,000+ implementations and ideas shaped by over thirty years of innovation. The business processes, embedded within their products, are considered best in class and the optimum way to run a business.

A new implementation of ERP software is the perfect opportunity to reset any stray business processing and implement what the ERP providers consider to be best in class. The challenge is the flexibility. Business change is hard for many organisations, and often project scoping sessions end up recreating what already exists in the incumbent system or process. Frustratingly for purists, these can all usually be achieved within an ERP build – just in a different way. Operating with a backward-facing mindset isn't going to move a business forward – and is something we need to protect against.

If we wanted to, we *could* make our shiny new ERP system brew a cappuccino every time a value over a certain limit was posted to a specific general ledger account. Cool huh? But *should* we make it do that? No.

Resisting the temptation of new functionality

In the three scenarios above, the challenge to an organisation is: what, in the processing of [insert process], is *so* unique to our organisation that it could not be replaced by a best practice 'out of the box' functionality? And does that uniqueness provide a strategic advantage to our business? If the answer to the latter part is yes, then build in the change – it's what makes you different and successful. However, in 99% of the cases, the answer is no and implementing a best practice is the better route.

We can stop all in-project creep and other scope changes by making it difficult to put change into the 'to be' model. This is a really simple concept but is often overlooked. We must justify the change and evaluate all changes together.

Start by listing out all of the deviations from the initial scope or solutions. Ask the supplier or subject matter expert to fully cost out the change (be sure to include all costs) and estimate the time the change will add. Once you know the facts and the full cost profile of the change, look at the alternatives. Broadly speaking there are two: accept the system as it is currently defined or accept a manual workaround. Cost out the impact of both scenarios. Now analyse the advantages and disadvantages of each scenario and ask the critical question: does the functionality give us a strategic advantage that we need to preserve? The next step is to present the recommendation to a leadership team and let them make the final decision on the change.

Clients usually start with an ideology of running with no change in scope. But reality is different. Once business processes are redesigned, everyone starts to get a little uncomfortable and gradually becomes more open to changing the system. It's at this point that senior leaders must be strong.

A CASE IN POINT

One client implemented a global system to manage sales and distribution. Through strong and robust leadership, they only accepted four changes to scope in the entire project and all four provided real strategic advantage to the organisation. The project was delivered quickly, on budget and fully embedded within the company within months of deployment. The total cost of ownership, because of the adherence to standard, was very small – enabling the client to take advantage of upgrades and new functionality quickly. This new functionality led to further competitive advantage and now they are dominating their sector.

How to cost a change

When costing a change, ensure the following are included in the cost:

Design: How much will it cost to design the change? Do you have all of the information? Will additional workshops be required?

Build: What is the build time for the change? How many people are involved? Have you considered internal and external costs?

Testing: What testing will be needed over and above what is already planned?

System integration: How will the change impact upstream and downstream processes or systems? How will it integrate with sub-modules?

Business change: What is the implication on the business for this change? How complicated is it? What training, above that already provisioned, might be required? What additional communication? Does any existing training need to be rewritten?

Documentation: How long will it take to document the change?

Total cost of ownership: What are the support costs involved after launch? How will future upgrades be impacted?

Reporting: What is the impact on reporting? Are any new reports required as a result of the change? Do any existing reports need to be adjusted?

Data and cutover: What is the impact on data migration? Is cutover likely to be extended?

Plans: What does this do to the overall project plan? What milestones need to be adjusted? What about the resource plan?

Contingency: Everyone wants to drop this to save money, but build in some additional resources for the unexpected. Changing a core element of functionality late in the design period is cumbersome, and it's likely something will have been missed. Work with 5–10% to be on the safe side.

You can download a free project costing template from www.limelightsolutions.co.uk/publications/runfast/resources

Governance and control

Governance and control are often cited as the largest time thief of a project. Many project decisions need to be approved by boards, steering committees and occasionally even the shareholder board. Some decisions do indeed require this route – fundamental changes to operating models, staffing levels, etc, but it's important to moderate so this doesn't become a bottleneck. Too many projects have stalled unnecessarily waiting for a decision.

At the start of any project, it's important to work with the senior team to define what decisions need to be made by the steering group and what level of delegation they are prepared to accept. The more decision-making responsibility that is pushed down to the project team the faster the project will execute.

Finding the right balance

Resolving the governance and control balance is a challenge – especially if the organisation is steeped in history or very 'command and control'-bound. There are some approaches that may help to break this deadlock and allow for decision-making delegation.

- Agree to run key decisions by one of the board representatives on an ad-hoc basis, and ensure you have access to them 24/7. This can be relatively informal – a breakfast is ideal, so no one's schedule is interrupted.

- Agree on a fast-track decision process for big decisions. Ideally electronic, with agreed SLAs. This should only involve the key decision makers – not the entire steering group.

- Agree to retrospectively review any decisions made over the previous week in the next steering committee meeting. This is predominantly for information sharing and ensuring open communication.

- Ensure the why, objectives, and anticipated benefits are so deeply ingrained in your team that any governance group will have complete faith in local decision making.

> **TOP TIP: Consider the role and responsibility matrix for the governance group created in the preparation phase. Ensure this contains the decision-making powers and explicitly state who is involved. These are the people you need to target for accelerated governance.**

Speed and feeling out of control

There is a perception that projects that run at pace feel out of control. This is probably one of the largest challenges leadership teams face. The answer, communication, is seldom recognised as the solution. Everyone on a project must communicate openly, frequently and robustly. People must talk. Encourage it.

Informal water cooler or coffee break discussions, formal updates, or even evening meals, are all essential to maintaining the breadth and depth of communication required. The typical challenge with a high pace project is the integration between different teams and parties. It's the responsibility of the entire team to ensure communication is flowing between all parties involved – both internal and external.

There are some great tools to aid in productive communication. Slack and Basecamp are two of the most common. We particularly like Basecamp due to the Twitter-style 'what have I worked on today?' update that all members receive from the entire team. It's great for spotting little nuggets of information that are key to other team members.

How much is enough? There are many studies on this, all drawing roughly the same conclusion: there is never enough communication. Ever. Tell people until they can recite your words back to you. Then tell them again. Several studies show the human brain remembers things more effectively through spaced repetition. Hermann Ebbinghaus set the precedent for this in 1885. In his paper *Memory: A Contribution to Experimental Psychology*, Ebbinghaus demonstrated that it is possible to dramatically improve learning by correctly spacing practice sessions or repetitions. This was further developed by Piotr Wozniak in the 1980s into an algorithm and computer-assisted remembering tool (www.supermemo.com).

Summary

It stands to reason that the faster the ancillary project processes, the faster a project will be delivered. Through the various techniques in this section, leaders must re-engineer their project processes for optimal efficiency. Governance, and indeed all processes, should add value – not steal time.

If we are able to define a project specifically from the outset (in terms of outcomes), establish a timescale that is highly aggressive (through optimal delivery and processes), and control the scope of delivery (through tight change control), we naturally have a controlled cost model for a project. Timescales and scope typically drive the cost, as the biggest expense is usually manpower. Control and processes are critical.

Reorganisation checklist

- ✔ We have looked at our processes and considered if they add value.
- ✔ I have optimised every process I can to ensure a timely execution.
- ✔ We have set a culture of continual improvement in our project, letting our experts do what they do best.

- ✔ We have a forum for ideas and suggestions for improvements, and a commitment to follow through with them.
- ✔ We have a robust challenge framework for changes to the core solution we are implementing.
- ✔ We are costing changes accurately.
- ✔ We have worked with our governance group and delegated as much control as possible.
- ✔ We have set up fast-track processes where board involvement is required.

CHAPTER FIVE

Outputs And Outcomes

Accelerated projects thrive when decisions are made at source. Failure to make a decision in a timely matter, or at all, burns time. If decision making happens swiftly, projects can become more agile and naturally accelerate.

> 'Procrastination is the thief of time.'
> —Edward Young

Outcome-based approach

The results chain is crucial to driving things forward in an accelerated project. Differentiating between layers helps drive decision making while allowing everyone to understand how their decision impacts the overall project.

An introduction to the results chain

The table below shows the results chain using a set of definitions originally developed by the Organisation for Economic Co-Operation and Development (OECD) in 2010.

The OECD results chain

Inputs	The financial, human and material resources used for the development of the intervention.
	↓
Activities	Actions are taken or work performed through which inputs, such as funds, technical assistance and other types of resources, are mobilised to produce specific outputs.
	↓
Outputs	The products, capital goods and services which result from a series of development interventions; may also include changes resulting from the intervention which are relevant to the achievement of outcomes.
	↓
Outcomes	The likely or achieved short-term and medium-term effects of an interventions outputs.
	↓
Impact	Positive and negative, primary and secondary long-term effects produced by a development intervention, directly or indirectly, intended or unintended.

In this results chain, inputs are used to carry out activities. Activities lead to services or products delivered (outputs). The outputs start to bring about change (outcomes), which contribute to the overall impact.

Potential confusion

Although in theory these different areas are easy to distinguish, in practice it can be quite difficult. There are three areas of overlap where there is often confusion.

There is sometimes confusion between activities and outputs. Some activities are clearly not outputs; for example, talking to users. When it gets to the level of 'designing a system' it's easier to see how there might be confusion. The act of 'designing the system' is clearly an activity, while the actual 'system design' is often considered an output as it's a product of a task. This confusion is common, and many projects feel unfairly treated when their outputs (or output indicators) are criticised for being too activity-based.

The second confusion is between outputs and outcomes, and here the difference is more subtle. The simplest way to distinguish the two is by recognising that outputs are tangible things that can be delivered, whereas outcomes are the cumulative impact of outputs that result in change.

The third point of confusion is between outcomes and impact, and here it is largely a matter of judgement.

In a desire for a less complicated life (see the previous chapter), we define outputs as the services or products delivered that are largely within the control of a project. Impacts are the lasting or significant changes in people's lives brought about by one or many interventions, and outcomes are everything in between.

In an accelerated project, the more decisions and actions made that are lower down the results chain, the faster the project will deliver its outcome and impact.

Considering the above, try to map out the results chain for your project. Describe the outcomes and impact you wish to achieve and turn them into a quantitative measure. Confirm that the outcomes are actually linked to your outputs or activities, implement the measures, and track them over time. This allows project teams to track, demonstrate and increase their success – this data will allow you to confidently and appropriately communicate your impact and value.

Decision making and empowerment

A decision that supports the results chain, and ultimately the why of the project, must be timely and for the good of the project. It is easy to be pulled into group thinking or to feel exposed when decision making. Sometimes the right decision is not the most popular, either within the project team or other corporate departments. It may invoke changes to processes or departments, or even result in changes to headcount.

Part of running an accelerated project is the recognition that decision making is best placed at the source of the expertise. This is typically lower down the organisational hierarchy within the project team: the project team are the subject matter experts.

In a traditional project, the decision making usually sits with senior leadership. If the project is implementing a new finance system, the best people to make decisions on its structure and operation are the operators of the system, not the leadership. It's highly unlikely that the senior finance lead will have even logged into the incumbent finance system (aside perhaps for reporting), let alone know how to operate it day-to-day. It's not their job, so how can they make decisions on behalf of the team? It makes far more sense for the daily users to make those decisions and the senior finance lead empowers them to do so (with the proviso that they can still get the information needed for their own role).

In the example above, the senior finance lead empowers their team members, or project representatives, to make the decisions for him/her. As the team are making decisions aligned to the why, the senior leads know they will align with the desired outcome. They push the decision making to where it matters and is most effective – down the hierarchy.

It should be noted that we are not pushing the accountability of the decision away, just the responsibility. If the financial system can't produce end of month or year reporting, the senior finance lead will still be accountable for the lack of compliance. Leadership must understand the difference.

TOP-HEAVY GOVERNANCE

Two clients, 'Client A' and 'Client B', are both similar in size, revenue, industry and complexity. Competitors, in fact. In working with Client A, they fully embraced the push of responsibility (not accountability) down to the project team. The team was fully engaged and motivated by their levels of responsibility. The strong why statement was their guiding light. Few decisions needed senior leadership approval and the project crossed the finishing line (yes, it was accelerated) with minimal organisational impact.

Running about six months behind, Client B executed a similar project. The culture of the company meant that senior leadership refused to let go of decision making and insisted everything be approved via steering groups and committees. Few decisions were made in a timely manner by the groups, and much was deferred to subsequent meetings. Project team members were frustrated with the lack of progress and indecision. Several of the project members requested to leave the project, and two actually left the company. The why statement Client B had didn't align with the style of execution and frustrated the project.

In the end, the outcome of each project was similar, but the cost to implement for Client B was three times the amount compared to Client A – and the duration of the project almost three times as long. Client A went on to further iterate their solution during the time Client B was still attempting to implement their project – increasing their market penetration. Client A is now a powerhouse in the industry, and Client B is struggling to make a dent.

Supporting decision makers

In order to push decision-making responsibility down the hierarchy, we need to support those making decisions and manage them accordingly.

Boundaries: There are some decisions or actions we might not want to push down the hierarchy – like those that impact the wellbeing of staff or headcount. These will differ per project, and the PMO (in conjunction with the senior leadership) may want to discuss these in the preparation phase. Set boundaries for the decision maker and make it clear what is, and is not, in their remit. The latter is likely more important as it gives the decision maker confidence in their scope.

Communication: Communicating the boundaries and the decision-making capability of the team is important. Everyone needs to know who can make and approve decisions.

Confidence and support: Decision making can be a daunting task. It's important to show confidence in the individual and support their decision once it's been made. Check in regularly with the decision maker, ensure they are comfortable with the level of responsibility they have, and let them know you are there to support them. Even if the decision maker is wrong and decisions need to be backtracked, help them deal with the consequences and give them your full support throughout. Senior managers and project leaders should offer themselves as a sounding board if required – a decision maker may want to walk through their options. If this is the case, allow them to explain the thought process to you, but empower them to make the final decision.

Training: Some individuals tasked with decision making may need training to be confident or successful. Depending on the culture of your organisation, this process might be a little unusual for some of the team.

Praise: Praise the team when they make a decision without the usual procrastination and debate. You will need to build a culture of fast and swift decision making to run an accelerated project, and encouragement goes a long way. Operating models and system design are generally items that are procrastinated over for weeks, if not months. If your team can conclude something within a single week, shout about it and communicate it freely. It will inspire others and lead to greater levels of confidence within the team.

Embracing failure

This section is split into two areas: the overall perception of failure, and the act of making a wrong decision and its impact on the individual and the team.

The perception of failure

> 'I have not failed, I have just found 10,000 ways that do not work.'
>
> —Thomas Edison

In the digital era, we need to move past the notion that failure is inherently negative. This notion has come around through years of black and white thinking. As children, we are taught right and wrong. We are rewarded for doing right and reprimanded for doing wrong. We learn from experimentation, yet become afraid of making mistakes.

Some of the greatest developments of our time exist because of failure: Playdough was originally invented in the 1930s to remove stains from wallpaper; 3M invented the Post-It note when it created a glue that didn't work. Considering both of those products went on to make millions for decades, failure can't be considered inherently negative.

There are some great companies that love failure. Indeed, many see failure as the route to perfection, knowing that much can be discovered along the way.

Toyota, the Japanese car manufacturer, is probably the greatest example of this. The teams at Toyota pride themselves on their production capabilities and methodologies. They run lean processes and methodologies throughout their build line and have a highly optimised design-to-manufacture process. On the production line, Toyota rewards their colleagues if they are able to stop their production processes – if someone can make a mistake in the process, the mistake needs to be protected against.

In order to run accelerated projects, we have to accept that decisions will be made that might not be right. In rapid develop and test cycles things may not work as we intend. Work to build a culture wherein the idea of failure is accepted, acknowledged and moved through while the corrective action is taken. There shouldn't be criticism or blame, just a diagnosis, acknowledgement and then realignment as the team works together to provide the corrective action. Remember, none of us are immune from mistakes.

Fail often, fail early

While acknowledging we need to create a culture of accepting failure, we need to try to protect against failing *late* in the delivery schedule. In accelerated projects, the later re-work is identified, the more risk and pressure we create on the timeline – so fail often and fail early. Failing early requires teams to think about *how* they structure their work. We often refer to this as early prototyping or the creation of a minimal viable product. The goal is to get something to the user community as quickly as possible to test concepts, ideas and logic. This allows for early critique and enough time for any re-work. We can even scrap it and start again if necessary. Using this approach, the project team can get their work in front of key users and stakeholders, assess the degree of fit, and then resolve or change the sticking points. The rough edges can then be tidied up, and the solution completed and released for general testing.

In the worst-case scenario, if something is fundamentally not going to work, we find out early enough to assess the impact and take the correct course of action.

Impact on individuals and teams

In accelerated projects, while we may create a culture that begins to accept failure, we must not lose sight of the impact failure has on individuals and the team. This acceptance may be a new concept for the organisation and it may be pertinent to remind teams and individuals that failure is acceptable. We need to ensure that those who have made decisions directly resulting in the failure are not feeling vulnerable or personally impacted. It is, after all, a team sport – and all of the pieces need to work together to achieve the end goal.

Saying no

Moving to an accelerated and outcome-oriented approach means moving decision making lower down the organisational or project hierarchy and encouraging failing often and early. Having set this framework up, one of the common objections we hear from team members is that they feel they can't say 'no' to their customers, peers, colleagues, managers and senior leaders.

The seven steps for saying 'no'

1. Ask for clarification

When someone challenges a decision, the first step is to communicate and develop a clear understanding. Do this by clarifying anything that may be vague and ask why they are disagreeing. They are providing you with the opportunity to ask for a detailed explanation. It's up to you to do something with the information they provide, but it's a win–win either way: they walk away pleased someone is listening and cares about their opinion, and you receive more information and flush out the reasoning.

Carolyn Kopprasch, Chief Happiness Officer at Buffer (what an amazing job title!), recommends asking what the individual is ultimately trying to achieve when they challenge a decision: 'We can learn a lot from those kinds of conversations if they're willing to indulge us.'

Going into details about any objections will help provide perspective – this holds true for good and bad ideas alike. Consider what the objection is trying to accomplish. When you understand the deeper issue, you may not give the person exactly what they asked for – but you are more likely to help them get the solution they need

2. Explain what's going to happen next

> 'At Eventbrite, customers will ask for niche features the company probably won't entertain,' says Vivian Chaves, a customer experience associate manager. 'Eventbrite support responds to those queries by letting the customer know that while there's no way to fulfil the request at this time, they'll share it with the product team, who will review it and scope it in relation to other initiatives. They also share how approved requests are placed on their roadmap, followed by coding and testing to ensure a smooth integration with their existing product.'

When presented with an objection, whether from a team member or customer, it's important to make them feel confident that their challenge will not disappear into the ether – and that there is a process for handling requests.

3. Be honest

People can spot insincerity, and dishonesty can come back to bite you. It's better to say no now and cause some disappointment instead of setting incorrect expectations and trying to manage them further down the line. If you know the answer is going to be no, then be honest upfront.

4. Reframe the 'no' using positive language

It's possible to convey a clear 'no' without ever using the word. Instead of 'no, we are not going to do that', try phrases like, 'I can see how that would be useful, but we don't have plans to add that functionality yet' or, 'while there is currently no way to do that, we appreciate you taking the time to let us know what you're looking for – thank you for reaching out.'

5. Make the person feel heard

People want to know they are being listened to and taken seriously. Small touches, like using the person's name, and understanding phrases go a long way. Thank people for raising objections. Whatever their issue, it was important enough for them to take time out of their day to contact you. Acknowledge the effort and your gratitude for it:

> 'We really appreciate the feedback, James. It's great to see you are so passionate about the new processes and I completely understand where you are coming from. The challenge we face is *x*, but why don't we look at trying to solve it together by doing *y*?'

TOP TIP: Reply quickly, but not too quickly. When you can modify your standard communication with the customer's name and an acknowledgement of their specific issue in thirty seconds, it can make some people wonder if their email even got read. It's okay to let non-urgent emails sit a few extra minutes.

6. Offer alternatives

You want your project team to be happy, not marginally satisfied. When you don't have an answer someone wants to hear, point them towards a workaround or even another solution. The resulting long-term loyalty will outweigh any short-term disgruntled behaviour. Be sure to be clear about any trade-offs – by presenting options with trade-offs, you elevate the discussion and no longer look like a project leader, but a collaborator. The best solution often won't be any of the alternatives you provide, but a compromise between many alternatives.

7. Explain the reasoning behind the current design

When people understand the rationale behind the existing design, they are more likely to be forgiving. Remember that people outside of the project team may not know details about a certain issue. Respect their lack of knowledge – they might be subject matter experts elsewhere.

Consider listing each problem and presenting the factors that make the request unfeasible. Present any counter argument in a concise and easy to understand manner. The next best thing to giving customers what they want is making them feel as though their ideas are taken seriously.

These techniques will strengthen your personal relationships as well as your reputation with the project team. Remember things will not always be ideal: even when presented with a vastly better solution, sometimes the other person won't move from their position. A stalemate is never a good outcome, so agree to disagree: sometimes it's the only way forward. There will be other opportunities and being gracious can buy a lot of goodwill.

How to structure meetings that matter

While this may seem a slightly odd section to include in an outcome-based chapter, there are too many examples of costly meetings that don't add value, don't come to a decision, or lack traction after a decision has been made. All of these work against the very nature of an accelerated project, and it ultimately comes down to how the meetings are being executed.

The three types of meetings

Broadly speaking, within a project, there are three types of meetings.

Informing meetings: These are the most straightforward meetings wherein one member, usually the chairperson, has factual information or a decision which affects all those present, which he/she wants to communicate.

Consulting meetings: These are meetings used to discuss a specific subject and can be useful to get participants' views of a policy or idea.

Problem-solving meetings: These meetings depend on the lead describing the problem as clearly as possible. Members should be selected according to their experience, expertise or interest, and then given as much information as possible to enable them to generate ideas, offer advice and reach conclusions.

Planning and preparing

The meeting lead is critical to the success of the meeting. They must plan, organise, and control the discussion of subjects on the agenda; encourage and develop harmonious relationships; motivate the individuals by encouraging all to contribute; reward effort; and support individuals in any difficulties. Before every meeting the lead should ask the following questions:

What is the purpose of the meeting? All meetings must have a purpose or goal. Don't have a meeting just for the sake of a meeting. Consider what you want to achieve and what outcome you are seeking.

Is a meeting appropriate? Consider the appropriateness of the meeting. Is another form of communication more appropriate? Unnecessary meetings waste time and are costly to an organisation. Only have meetings where they are required.

How should the meeting be planned? Take some time to plan every meeting, no matter how low-level or informal. Planning will help dictate the content, and ensure the right people are present.

Who should attend the meeting? Where meetings are required, ensure all the right people are in the room. Deferring to secondary meetings wastes time – the Achilles' heel in accelerated projects. If the meeting outcome is a decision, ensure all the key attendees – especially the potential opponents of the decision – are present. The outcome is the fundamental focus in accelerated projects. If key members can't attend, reschedule or ensure they have delegated their responsibility to someone who can.

What preparation is required for the meeting? Circulate the agenda prior to the meeting and include any briefing papers in the invitation – but *do not* spend the meeting reading the paperwork. Make it clear from the outset you expect the material to have been read and understood before the meeting begins.

During the meeting

Running an effective meeting is a skill, and there are numerous courses and books on facilitating great meetings. Some important takeaways:

- Be clear about the objectives of your meeting, and don't allow other topics to creep on to the agenda.
- Close down any discussion that does not contribute to the outcome you are looking for.
- Start and finish on time. Set the expectation of doing so at the start and usher the meeting forward to meet the set timeframe.
- Spend two minutes at the start of the meeting framing the context. It sets the scene and gives meeting attendees a few minutes to consider their thoughts and get into the right mindset for the meeting.
- Introduce anyone new to the group. If it's a regular meeting only do this once, but ensure you introduce new attendees going forward.
- Stick to the agenda and work through items in chronological order to their conclusion.
- Make a note if anything important is raised but not relevant to the particular outcome. Move the conversation on and agree to return to it in a more suitable forum.

- Connect with your attendees. Use humour where appropriate and align with your team members to help build deep relationships.

- Agree on an action plan for any issues and seek confirmation of decisions made. Give everyone an opportunity to respond to the decision.

- Set a timer and stop when 10% of the time remains. Use the remaining time to review the action items, set the agenda for the next meeting (if required), and get aligned on communication messages. Discipline at the end of your meetings will support better execution.

After the meeting

Once a meeting has finished, act on the following points. They will add value and ensure follow-through on the decisions and outcomes.

- **Reflect:** Analyse how the meeting went, including any feedback from the group, and identify learnings for next time. Even the most experienced leaders have room to learn.

- **Celebrate:** Celebrate any significant progress made in the meeting. This might have happened already, but chances are you can do more to acknowledge it – celebrating motivates more of the same behaviour.

- **Transcribe decisions and actions:** Write up your notes, focusing on the outcomes and actions. Be sure to include who agreed to do what, and by what date.

- **Distribute meeting notes:** Send a copy of your notes to all who were present, and those who missed the meeting. Include a reminder about the next meeting, if required.

Summary

Accelerated projects require clarity. Encouraging your project team to focus on outcomes, and the factors that directly impact them, is crucial. Empower your project team to make decisions – the faster a decision can be made the better. Support them in this process. Ensure teams are competent and capable of controlling their area through robust meetings and saying 'no' where appropriate. Everyone contributes to running things at speed.

Outcomes checklist

- ✔ My project has a results chain that makes sense and is cohesive.
- ✔ My team understands the results chain and the need to make decisions that get the project to the outcome faster.
- ✔ Appropriate decision-making power has been given to the project team.
- ✔ The project has considered support for decision makers.
- ✔ The project has an approach to deal with the perception of failure and support for the individuals involved.
- ✔ My team understand the need to say 'no' to certain requests and feels comfortable doing so.
- ✔ When meetings are required, my project team knows how to ensure they are productive and efficient.

CHAPTER SIX

Mindset

Changing organisational mindset, or that of individuals, is probably one of the largest hurdles any business leader will face. Organisations, especially large and traditional ones, tend to be like oil tankers. They will turn, but it takes incredible effort and a lot of time. Individuals are often easier to change, but their external environment will influence them heavily – and without results or reinforcement experienced first-hand, it's easy to fall back into old ways. People tend to reflect the organisation in which they work, bringing along habits and behaviours from previous experiences throughout their career.

Organisational culture

Before influencing the mindset around a project, it's important to recognise the existing culture of the organisation and the team around you. Individuals are shaped by the organisation, and the organisation is shaped by the past – so look to past projects, initiatives and changes to gauge the maturity of the organisational culture. The more forward-thinking the organisation, the more change they will be prepared to attempt, even in the face of challenges. Organisations with a long, bureaucratic history often prefer to look backward at their heritage and what they have achieved, rather than looking forward to the future.

A forward-looking, change-friendly organisation often already has the mindset you need to run an accelerated project. These organisations will likely have completed something similar in the past and will embrace the challenge. The people within these organisations tend to reflect the positive, can-do attitude required of the delivery team. The organisational mindset and culture, in this instance, can simply be adopted as a pre-packaged asset for the implementation team.

Those organisations that haven't yet embarked on a significant change project or are generally quite adverse to new ideas will need their culture and mindset managed. They will need focus, effort and attention to ensure the team remains committed to the goals of the accelerated project.

There are three predominant points to remember when it comes to culture and mindset:

Tackle it early

The organisation's culture and mindset need to be identified early – preferably before the project has started. It should form part of the consideration of the required timescales, team, tasks and the operating model of the project. If intervention is required, immediately start addressing it. From the moment the project is underway, effort should be spent attempting to create a mindset or cultural shift. The project will be influenced by the early routines and patterns it adopts. Ensure these are aligned with how you intend to run the project.

Tackle it head on

Be assertive around dealing with organisational culture and mindset. It will require direct communication, and actions you take will almost certainly ruffle feathers. Lean on the backing of the executive leadership for their endorsement and be relentless in the actions you take. You will need to demonstrate to the organisation the new operating model for the project, and secure buy-in from all key stakeholders. Communicate continually throughout this phase, as it will be unsettling for some and you will need to win over the nay-sayers.

Continue to address challenges

Remember you are dealing with an oil tanker – and a single tanker takes up to forty-five minutes to complete a 180-degree turn, and twenty minutes to come to a complete stop. If the tanker pauses or changes direction midway through, everything is exacerbated. The point here is to never stop – ensure there is constant continual change towards the goal. When distractions arise, always maintain a handle on mindset and culture. They will drive behaviours and ultimately influence the project more than any tactical intervention. Even if things are positive and progress has been made, keep working on the mindset and culture – it's tremendously easy for individuals and the team to fall into old habits.

Cultivating a can-do attitude

Having a winning attitude is key for accelerated project delivery. The team has significant hurdles to be addressed – these will vary from project to project, but with reduced timescales, all of them will be exacerbated compared to a normal delivery. It's important to be confident in the seemingly impossible being achieved and communicate that belief with passion. This is done by setting goals, addressing challenges, creating camaraderie, promoting creative thinking and supporting and leading the team to ensure nothing slows down the pace of the project.

At the beginning of the project, meet with the team to discuss what is likely to slow down delivery – then resolve those threats aggressively. If the perceived issue is business engagement, then meet with heads of departments and confirm their commitment and resources. If it's internal change processing, then work out ways to accelerate the process – but always maintain integrity. These issues, and their timely resolution, will demonstrate to the team that things are changing to support them in their goal: the delivery of the why. It will demonstrate the pace the project requires.

These sessions should ideally be run at the start of a project, in the preparation phase. Get the issues on the table, then remove them as quickly as possible.

COLLECTIVE PROBLEM SOLVING

In working with a particular client, we orchestrated what became known as weekly battle sessions – perhaps an indicator of the culture of the client. During this session, everyone was encouraged to raise what they felt was holding them up – regardless of their role or position within the organisation or project. The goal was to solve the problem during the session, with minimal to no follow-up actions. This kept the sessions lean and aggressive, and ensured the right attendance and oversight at the right level of the organisation. The result was a mindset shift – and gradually the organisation moved from a 'cannot' business to a 'can' business. Eventually the obstacles were removed outside the formal sessions as part of the regular flow of work. This made a huge impact on the project team's productivity, speed to resolution was increased on all aspects of the project, and the solution delivered was a better fit – the business and the project team were aligned. Change management was also easier to deliver, as the existing barriers had been removed during the project processes.

Speed of resolution is really important here. The faster the project leadership can react to the issues being raised by the team, the faster the team will respond in accordance. By doing this leadership can show their support and engagement to the project – it sets a pace for the future. The team will hit future hurdles and know they will be resolved quickly – no issue becomes too big for the team to address.

Pushing the team

The project lead should set aggressive targets for teams and the individuals within them. People work within their comfort zones – it's just human nature. Ask them to give 50% more effort and they will break; ask for 10–20% and you'll see dramatic improvement. Learn how hard to push your team – then learn how long to back off to allow them to recover. The delivery schedule should be based on a push, rest, push, rest cycle. No one can run at 120% for any length of time. They can, however, for a short while or sprint. Excellence is always in the margins.

When the team delivers something remarkable, shout about it. Working at 120% is incredibly difficult, emotionally and physically, so show enthusiastic appreciation to everyone when the work package is complete and things calm down. Reward the team, and reward their families. Go beyond usual thinking and find something everyone will aspire to receive. It will further motivate and inspire.

The key to this section is to learn how to show everyone in the organisation that you can and are winning the project delivery challenges. The seemingly impossible is looking increasingly like the very plausible.

Motivation in the digital era

The section above purposely ventured into motivation – creating an unbreakable can-do attitude is key to running an accelerated project, and motivating team members is crucial. All leaders should aspire to become a master of motivation, as it's the most powerful tool you have. However, everyone must also be aware of the subtleties of motivation across generations. Motivation during the baby boomer period is very different to that of the digital era; just like that of Millennials and Generation Z. Be conscious of any training or business help books you refer to and the era they were written in.

The fundamentals are still relevant, but the nature of work has changed and the new worker generation have different expectations, values and aspirations.

One of the increasingly common trends to be mindful of is social and work–life balance. Previous generations tended to operate a work day/social day split. They would work from 8am to 17:30pm, for example, and then spend the rest of their time socially. This boundary is gradually fading. New generation workers are working on a getting-the-job-done basis, and want to enjoy the flexibility of a two-hour social lunch break, making up the workload in the evening. Or they may work a Saturday morning and not work a day during the typical work week. These workers are always on and will read emails out of office hours, replying immediately when necessary. On the flip side, they also expect to be able to check social media or message friends during office time.

Flexibility and motivation go hand-in-hand. Don't try to fight against the trends of any era – go with them and use them to your advantage. Watch for people who will try to take advantage, but it's important to respect the culture of work in the digital era.

Another key trend worth considering is that of social responsibility. There is a lot of resonance, especially among the Millennial and Generation Z population, around social responsibility and improving the world in which we live. Companies and projects that have this as a core pillar often find a better fit with these workers. Work out a way to build a social aspect into your project – the more creative the better. As a consulting practice, we educate children in some of the world's poorest areas through a direct link to our billing days. For every day we bill, we educate a child for a month. It's our way of giving back, and links to the United Nations Global Goals framework (www.globalgoals.org).

Work in the digital era

The nature of work has changed in the digital era. We now operate with far more flexibility, and will almost certainly see a further movement towards the 'getting the work done whenever/wherever' philosophy. Teams should expect to be distracted – the 'nudge-based' world we now occupy constantly pulls attention away from tasks. Phones buzz every few seconds, social media demands updates, email marketing is prolific and the constant consumption of knowledge through blogs and articles is commonplace.

The digital era is the most chaotic in terms of sensory or data overload. We have never been so inundated with alerts, pings, alarms and nudges from different technology platforms all competing for our time. Managing these trends doesn't mean simply ignoring or trying to fight them – instead we need to leverage these factors, appreciate the quirks and generate value from them. If distraction is a threat to accelerated projects, then using the same tools causing the distractions during project delivery will help keep teams on track. Set up a Slack channel, tweet about progress or use Basecamp to align teams. There are so many tools out there; pick one that suits your project delivery and use it to your advantage.

REAL-TIME COMMUNICATIONS

One of our clients opted to use Twitter as their communications platform. The team subscribed, with their personal accounts, to the corporate project Twitter handle – and every time something of significance occurred, the team published a tweet about it. Everyone on every team was encouraged to tweet, with prizes being handed out for the most tweets or contributions. The result was superb communication within the project, and the stakeholder team could get a live picture of what was happening on a daily basis. Bad and good news alike, led by the project leads, was published onto the platform. As confidence within the team grew, it became an incredibly powerful 360-degree communication channel. Even the CEO joined in on some of the conversations and provided valuable insight into certain problems. This added immense value and further motivated the teams to push delivery as fast as they could.

Customer	*Great design workshop today. Really enjoyed the session and it provided a lot of clarity for us... Looking forward to getting started*
Supplier	*Ditto – we felt it was a good session. We have everything we need to create the straw man – see everyone in 5 days*
Customer	*Something I forgot to mention... we have a subtly difference process in this business unit – can we meet tomorrow?*
Customer	*No problem – will get something in the diary*
Customer CFO	*Remember the goal ☺... standard processes – let's see what we can do about this without changing things*

Cross-team problem solving and mindset

One of the main obstacles to overcome during an accelerated project is silos. It's easy for teams to be overly focused on their own role, especially if you are delivering an enterprise-wide project. The larger the project, or larger the project backlog, the more likely team members will sit with blinkers on, each immersed in their own challenges. This is a problem waiting to happen. To deliver an accelerated project all components must move forward simultaneously, not linearly. If most of the project is at 60%, but one area is still at 30%, then the whole project is still at 30%. You are only as complete as your weakest component. Forget averages – they are misrepresentative and designed to make the project appear further along than it actually is. If one critical component is locked at 30%, your entire delivery is locked at 30%: 99% of a product isn't a product. Only 100% is. As leaders we need to ensure all components are moving forward simultaneously towards the goal, not locked in a sequential linear delivery pattern.

It's beneficial to explain this concept to your team, as they must operate as a unit to help and support the slowest or most problematic component. Problems need to be rallied around, group ideas considered, and tasks allocated and ultimately resolved. All areas must make traction together.

One technique to help foster joined-up teams is to look at holistic or big problems during formal problem-solving events. Take the team offsite, away from distractions, and focus for the day on resolving one particular problem holding up a workstream. It's important to actually take steps during the session to solve the problem, not just come up with ideas or theories. Work on the problem all day, regardless of who owns the problem or workstream – get everyone involved. On an accelerated project, the team must always move forward. Discussion needs to be minimal and action abundant.

Discussion vs momentum

Moving people out of discussion and into action is crucial to the success of an accelerated project. A discussion is ultimately procrastination and a hindrance to challenging timelines. This obstacle can be present at all levels of the organisation, from delivery teams through to senior management. People naturally want to minimise risk and want to know all the facts before making a decision. Unfortunately, by gathering all seemingly relevant facts the project stalls.

To overcome this, we need to get comfortable making decisions without knowing or understanding absolutely everything. Teams need to be empowered to act in the best interest of the organisation, led from the specific why statement created at the outset of the project. Decision-making responsibilities should be pushed down the organisational or project hierarchy and teams should implement prototypes or framework solutions to mitigate risk. There needs to be an attitude of trial and error. What must be avoided is spending weeks discussing before trying anything. Do and test, do and test. That should be the mantra of the team.

This approach has benefits aside from preserving the timeline. Creating several prototypes prior to finding the solution allows *what if* questions to be answered along the way. It shows senior teams the options that have been considered and why they were discarded. It allows innovation to flourish and ideas and solutions to surface. The maturity of a team working with this strategy generates confidence from the organisation and leadership team: the team will solve the problem, somehow.

We must remember the ultimate goal of an accelerated project – to deliver the why statement in the fastest timeframe possible. Keeping everything moving forward at all points is fundamental to this.

GENERATING SOLUTIONS

We encouraged a large client to run Problem #1 workshops on all of their projects. These sessions were designed to tackle the biggest problem on the project (and no others). They drilled down to the root cause (using the 5 Whys framework) and solved the problem in as many ways as possible – executing and testing proposed solutions live in the workshop. Some solutions were elegant, others hideous. Some created business process issues and bottlenecks, others allowed the organisation to shed resources. Many of the ideas suggested hadn't been considered previously. The client adopted rapid solutioning and the 'fail fast' mentality. They were able to evaluate options and decide on an appropriate solution quickly. They knew it was the right solution because they had already evaluated many alternative options. Within three hours, the biggest problem on the project had been resolved.

Pessimists and how to deal with them

There will always be pessimists on your project convinced that something (or many things) can't be done. They are often steeped in the thinking of the past and can't see a different way forward – sometimes through no fault of their own. Getting the most from each member of your team is key, and dealing with a particularly pessimistic colleague can be exhausting. However, you need to get the most from everyone, and converting them must be a high priority before the pessimism spreads to other teammates. Getting results from pessimistic colleagues can be incredibly rewarding, and the effects tend to ripple across the entire team.

Addressing negativity

Ignoring negative comments and hoping they will go away is dangerous. Your pessimist colleague is likely pessimistic because they feel they have been ignored or devalued by others. Whether this is you, someone else at the company, or people they worked with previously, not addressing their negative feelings can exacerbate whatever issues they have. Make sure to have regular conversations with all members of your team, especially the ones who are pessimistic about certain aspects of the project or company life. Take them seriously and develop ways you can change what is upsetting them.

Ask for details

If your pessimist is vocal during team meetings or creative sessions, don't respond negatively. An eye roll or attempted diversion will only exacerbate the issue and create a larger gap within the team. Ask them to expand on their point and use it as a catalyst to start a conversation with the other people involved. This will help them feel less isolated from the team, and confident that you are taking their concerns seriously and want to address them in a practical way.

Asking a pessimist to expand on their viewpoint also takes away some of the shock value that many pessimists get from saying a negative comment. Not all colleagues with pessimistic attitudes revel in the attention but normalising the behaviour will take away the entertainment value. If they are just saying things to be funny or get attention, but don't really have a point to back it up, they will look foolish and are unlikely to continue with the behaviour.

Re-evaluate your team

Try looking at the people they are working closest with. Are they just as negative? If so, try switching up the team dynamics and see if this results in a change in their behaviour. Having them work alongside a motivated employee may be enough to encourage them to be more engaged. If this doesn't work, you may need to consider moving them to another team.

Make sure you monitor the other members of your team closely to ensure your pessimist is not having a negative effect on them. The last thing you want is for their pessimism to bring down the rest of the team. A consistently pessimistic colleague can be damaging to overall team morale – so it's important to recognise where to draw the line.

Summary

The mindset of team members in an organisation is critical to running an accelerated project. Organisational heritage and history play a large part in shaping both the individuals that will join your project and the stakeholders your project engages with. Analyse the impact of this early and take appropriate action to ensure the overall mindset works for your project, not against it. Tackle any issues head on. Leaving things to run on their own will almost certainly result in unnecessary delay. Be persistent and reiterate messaging throughout the course of the project.

The importance of fostering a can-do culture must not be underestimated. Your delivery team should feel unstoppable, and you should do everything in your power to facilitate this. Motivate your teams using things that are relevant to the demographics of the particular generation of each individual. Remember the new workforce demographics have different values to those of previous eras. Work with these values, find tooling that works for everyone, and don't try to fight the underlying concepts of the digital era. It will frustrate everyone and work against the principles of an accelerated project.

Inevitably, you'll have people on your team who believe things won't work out. It's critical to manage these individuals and control their impact on the rest of the project team. The popular wartime phrase 'loose lips sink ships' is highly relevant to the success of the accelerated project.

Mindset checklist

- ✔ We recognise whether the existing organisational culture will assist or hinder running an accelerated project.
- ✔ We are tackling culture and mindset issues early and head on.
- ✔ We have fostered a can-do attitude with the delivery team. They feel invincible!
- ✔ We understand that different generations of workers have different expectations and motivations.
- ✔ We understand the nature of work in the digital era and the impact it will have on our project delivery.
- ✔ Our teams are highly tuned into group problems and problem-solve as a unit, helping the weakest at all costs.
- ✔ We understand how to manage a pessimist and the need to do so expediently.

CHAPTER SEVEN

People

Stating that assembling the right team is central to success seems as obvious as telling people to increase revenue to grow their business. However, it's important to understand the value of having a great team of colleagues behind you, all working towards a common accelerated goal.

Executive commitment and governance

To ensure the successful implementation of any project, especially one operating at a high velocity, it's essential for the project to have strong support from senior management. Projects lacking senior support face an uphill battle in receiving business-wide buy-in and adoption. When everyone in the organisation recognises senior team buy-in, you have a much better chance of everyone embracing your new system and transitioning to a new operating model – developing accompanying policies and procedures along the way.

If everyone in the organisation can see the board, executives and steering group leading from the top, everyone else is more likely to follow suit. There is a note of caution here: middle management often find themselves in the impossible position of trying to keep both senior management and operational teams happy, so they must also have a voice in programme governance. These are some of the most difficult yet important people to get on board.

The team

Implementing an accelerated project requires an experienced leader and team. Many organisations try to cut costs by getting their project to fit around existing responsibilities, but this will ultimately lead to individuals burning out and the project failing. By backfilling the required positions, you will save time and money, and prevent excessive stress on your team.

If you are working with a third-party consultancy or software partner, they may also bring in consultants to guide the business through implementation. External consultants will probably form a large portion of the budget, so ensure they are productive and can work unhindered. Making them feel like part of the team is essential, so take good care of them: be sure they know where everything is, understand all of the processes, and treat them like one of your own. Close collaboration with this third-party workforce will be key to your success.

Talent

You will never succeed if you don't have the right people on board. Accelerated projects, and in particular their change aspects, can't implement themselves. As with project success in general, it begins and ends with leadership. Accelerated projects aren't easy – they push people out of their comfort zone, run at a pace many aren't comfortable with, and work amid a high level of ambiguity. Putting the organisation's best people in key project roles ensures a solid project and sends a message to the organisation regarding the seriousness of the outcome. Longer term, these people become ambassadors for their work and champion it for years to come, often as they progress through management ranks. They are fully committed from the outset and understand the value the system brings. The talent an organisation assigns to its team demonstrates to a large degree its commitment to and comprehension of long-term organisational health.

Leadership

Leading an accelerated project requires working among many unknowns and much ambiguity. Leaders need to be comfortable with not having all of the answers and putting their trust in their team. They need to be skilled at facilitating and creating the desire to work, and then stepping back as team members rise to the occasion. Project leaders must be able to articulate the project why statement and objectives in a clear and concise manner, tailoring it to different audiences and motivating others along the way.

The more experienced the leader the better, but be sure to select someone who is competent in project management without being too rigid in their processes. Accelerated projects are not the same as standard projects: speed is the order of the day and that leads to ambiguity, mistakes, innovation and sometimes a lack of clarity and structure. Command and control does not work. Flexibility rules.

At some point, something on the programme will go wrong. Managers and leaders who can keep a cool head under pressure and help the rest of the organisation stay calm are critical, so ensure your leader has an unflappable nature with proven problem-solving skills.

Leaders don't necessarily need to understand the technology they are implementing – all they need to do is to get the most from their team, inspiring and coaching them where necessary. Much of the skills required are around human interaction. As team members are pushed to deliver to the accelerated timescales, the leader is in place to facilitate and remove anything that is stopping them from performing at their best. The goal is to get to get the team to peak performance – and then keep them there.

Leaders need to remember that individuals are motivated by autonomy. *Drive* by Daniel Pink is a great read on human motivation. He provides overwhelming evidence on this and lays down the challenge that we must all 'create the conditions for action – then move out of the way'.[9]

Removing barriers

One of the vital roles of the leader and executive team is to remove roadblocks for the project team. The project will almost certainly hit bumps and occasionally stall. This is the time for leadership and the senior team to rally and remove whatever is stopping the project from progressing. This is no different to Toyota reacting within seconds if the build line stops – everyone must respond. It's a crisis that needs instant attention. The sooner this is understood by the executive team the better.

9 Pink, D (2011) *Drive*, Edinburgh, Canongate Books.

HARNESSING COMPETITION

One of the best examples of this was on a major ERP project. The resolution of barriers was set up as a competition between all of the executive leaders. The project lead would send an email to all of them with the problem, and the first one to solve the problem was treated to lunch by the others. Simple but effective: and it appealed to their competitive nature.

Articulate the need for change

Before beginning any accelerated project, it is crucial to ensure that the entire organisation understands the reasons and strategy behind the desired outcomes. If decision makers don't support the need for change, or if project members and end users don't understand the outputs or outcomes, the project will stall or fail.

Articulate and endorse the desired outcomes. Spend time explaining why the change is needed and what will happen if things don't improve. Be firm and treat everyone with respect. Ask questions and actively listen to problems or concerns. Articulate the outcome of the project vision, explain why the change is necessary for the health of the organisation, and ensure this message is communicated effectively to everyone.

Humans are creatures of habit and change is often resisted. This is even more true when colleagues have been in a role for a long time and find new ways of doing things intimidating. Before you start an accelerated project, you must get everyone – your team, governance groups and stakeholders – on board. The entire organisation will need to understand the rationale behind the project and why it will be positive for everyone involved.

Know your history

I can't emphasise enough how important it is to understand the established business culture and how, historically, the business has responded to change.

Even the most successful business leaders make mistakes – but they learn from each mistake and use this knowledge to shape what they do in the future. Knowing the history of your organisation and what has and hasn't worked will be of huge benefit when it comes to implementing an accelerated project. In accelerated projects, there is no time to make these mistakes again. They must be addressed from the outset with a new and better way of thinking and working.

In order for accelerated projects to function effectively, you should consider how resulting changes will fit in with or need to be adapted to the culture of the organisation. Culture plays a large role in delivering success. The more you can learn from the past, good or bad, the better your chances of success.

Expectations of your team

One of the most important onboarding elements for an accelerated project is ensuring everyone understands its process and velocity. Ignoring this will cause a misalignment of expectations and timescales. Your team will need to be coached on how the project will execute, what is expected of them in their role, what the delivery timetable looks like, and what decision-making authority they have been delegated. The team must be comfortable with the approach and understand the demands that will be placed on them during the project. It will almost certainly impact or infringe on family life, so make this clear from the beginning. Explain the reward for being a part of the project and encourage team members to discuss the project demands with family members. The less unexpected tension the project causes, both at home and at work, the better.

Decision making is key in accelerated projects. Ensure your team fully understands what has been delegated to them and that they are comfortable with their responsibilities. As per the previous chapter, provide assistance and support frameworks to aid team members through this. Don't leave them isolated, and make sure to check in with them on a regular basis.

Creating a successful praise and reward framework

We strongly encourage putting a reward framework in place for your project. Running an accelerated project puts a lot of pressure on individuals and teams. The GSD (get stuff done) motto is prevalent throughout, and there are often late nights, early mornings and weekend working. The additional stress and demands put on your team needs to be recognised, as does the infringement on family life. It's probably not what colleagues signed up for when they joined the organisation. Work out a metric to reward team members – perhaps linked to major project stages or milestones. Communicate the reward at the start of the project and ensure you follow through on your promises. It doesn't always have to be financial – some of the best projects reward team members with nights out, meals, trips or events. These are fun and encourage team bonding and collaboration.

Getting people engaged in the delivery of the project is essential both for the success of the project and for the overall sense of morale in the organisation. The more enthusiasm you can cascade through your team the better.

Home life impact

The best project leaders recognise the impact a major project has on family life, as there will almost certainly be extended working hours. Make it known from the outset that you understand the family pressure this adds, and that their commitment is appreciated. Ensure team members discuss the likely impact with family members in the preparation phase. Accelerated projects often require long hours – which isn't always possible given each individual's unique situation. It's better to identify any potential issues before the project gets underway.

Be sure to personally thank family members who are indirectly impacted by your project. This is a fantastic morale boost and shows the organisation cares. Trips to theme parks, carnivals, the beach – even a hamper sent home are all great gestures of thanks. Team members with an understanding and supportive family give the project it's best chance at success. Make it known how much the organisation understands and appreciates their behind-the-scenes contribution to the project's success.

Throughout the course of the project, ensure team members are comfortable coming to you if the project is creating difficulties at home. Everyone has to prioritise family. Respect this and help the team member through the challenge – they may need a break for a few days or to flip out of the role entirely. Either way, the project has a duty to the colleague through what issues it may have caused. Be sure to build an ethos of acceptance if this is the case. Team members should not feel guilty if they choose to put family first. A happy home life creates a happy project member.

TEAM BONDING

While working with one client, we took the entire team and their families to a popular theme park for a day out. We arranged coaches, lunch, and even gave each family spending money. Throughout the day we held competitions, for adults and children alike, around the park. And there was only one rule: no conversations about work! This outing encouraged everyone to get to know each other personally while providing a much-needed break from the project. The total event probably cost about £7,500, but the return the team gave back to the project was worth fifty times the amount. Appreciate your team and their families and show that appreciation readily.

Bulletproof communication

There are two types of communication on a project: internal (those within the project community) and external (project stakeholders, including other organisations, partners, government agencies and third-party providers).

Internal communications

Communicate as often as possible to the project community. In accelerated projects, things change quickly and it's important for all internal teams to understand the timing of the project, what is ahead, (and behind), and how their output integrates with other teams or individuals. In an Agile project framework, a daily stand-up is used to keep everyone abreast of progress and to highlight blockers. This works well with small teams in the same time zone: a great alternative for larger teams or international projects is built into a piece of software called Basecamp. It encourages broadcasting a Twitter-style update at a frequency the leadership team dictates, usually at the end of the work day. This keeps everyone updated with a quick scan of what everyone is working on, triggering discussion and collaboration in turn.

External communications

There is a golden rule for external communications: always be first. Good news or bad, ensure the project team communicates immediately and directly to stakeholders before they find out via any informal channels. Rumour and gossip networks have a fantastic habit of spreading bad news quickly. In this case, by the time a formal communication has been released, the damage has already been done. No news is not good news; it will only breed suspicion. A simple 'we haven't made a decision on x but are looking into it' message addresses the issue head on. These messages won't be read thoroughly by everyone, but no one will be able to accuse you of not keeping everyone informed.

Be sure to keep third-party companies up to date on the progress of your project as well. Your organisation will have spent time building these relationships – respect them and treat them as you would your own team. They play an important part in your value chain.

Dealing with complications

Conflict

Projects that run at pace are going to generate conflict. The majority of this misalignment and conflict comes when two teams have to interlock: what was agreed to informally three weeks ago may not now be the case. Change, or speed of change, may be causing a problem.

The leadership team needs to diffuse any tension immediately and focus the parties on coming to a resolution. Blame and resentment hurt morale – and timescales. The skilful leader needs to acknowledge the mistake, move through the emotional stage, and get the team working on the solution as soon as possible. Some of the best leaders I have met use phrases like: 'I am really pleased we found this misalignment with enough time to do something about it. These things are inevitable given what we are achieving. Good job for highlighting it. Have we thought about a solution as yet? What can I do to help?'

Mistakes

While delivering accelerated projects, you will undoubtedly face unexpected issues and your team will make mistakes along the way. It's a guarantee with the pace everyone is working at. These have the potential to seriously compromise a project if not managed effectively. If teams depend on business processes that were incorrectly designed, or face issues never anticipated or accounted for, there is a risk of reverting to wasteful workarounds at go-live.

When mistakes and issues arise, avoid the command and control reaction. Let the team resolve the problem as they see fit and work through to the solution together. The team always knows best. Provide support if they need it and facilitate any joined-up discussions that need to happen. Show them you have full confidence in them and empower them to lead discussions on the issue in meetings – their confidence will lead to a better, more considered solution.

Remember to never isolate individuals, or infer or allow blame – ensure you're supporting the solution *and* the team member(s) responsible for the mistake. Blame will only pollute the team atmosphere; you need everyone firing on all cylinders. Simply acknowledge a mistake has been made by the project (everyone is responsible for the deliverables) and start working towards a solution.

Getting the most from self-managed teams

Accelerated projects are at their best when the teams delivering the outcomes run themselves. Known as self-managing teams, they work to a milestone-based skeleton plan – how they achieve the milestone is up to them. The milestones are typically situated at major points, where functionality converges and teams are required to come together – like for integration development or test events.

Self-managed teams are incredibly motivating and empowering. They are typically 15–20% more productive than standard teams and generate far superior customer satisfaction. Commitment, innovation and creativity increase significantly. This approach aligns to the leadership principles we discussed earlier in this section: set the course and then get out of the way. Empowered to deliver, they will naturally push for the finish line, working together in doing so. They naturally remove any hurdles – for example, providing cover for one another's leave, ensuring nothing stalls in the process.

To get the most out of a self-managed team, set clear goals and ensure everyone capitalises on the diverse experience of their team members. Don't nominate a leader – leadership is shared, and the most experienced member of the current task will naturally assume the role.

One thing to look out for with self-managed teams is groupthink. Wikipedia defines this as 'the psychological phenomenon that occurs within a group of people in which the desire for harmony or conformity in the group results in an irrational or dysfunctional outcome.'

In other words, people tend not to speak out against something they feel is the wrong decision because they want to maintain harmony within the group. This is in fact herd mentality, and is characterised by eight core symptoms:[10]

- An illusion of invulnerability: members ignore danger, take extreme risks and are overly optimistic.
- Collective rationalisation: members ignore, discredit and explain away warnings contrary to group thinking.

10 Montier, J (2010) *The Little Book of Behavioral Investing*, New Jersey, John Wiley & Sons.

- Belief in inherent morality: members believe unquestionably in the inherent morality of the group, ignoring the ethical and moral consequences of their decisions.

- Stereotyped views of out-groups: the group constructs negative stereotypes of rivals outside the group.

- Direct pressure on dissenters: members pressure any in the group who express arguments against the group's stereotypes, illusions or commitments, viewing such opposition as disloyalty.

- Self-censorship: members withhold their dissenting views and counter-arguments.

- Illusion of unanimity: members perceive falsely that everyone agrees with the group's decision; silence is seen as consent.

- 'Mind guards' are appointed: some members appoint themselves to the role of protecting the group from adverse information that might threaten group complacency.

In summary, there is a risk that bad, or wrong, decisions get made. Educate the team at the start of the process and give them pointers to avoiding groupthink. These may include:[11]

11 www.highfive.com/blog/8-steps-to-avoid-groupthink-in-meetings

- Encouraging everyone in the group to evaluate ideas critically
- If you're leading the group, keeping quiet about your own opinions
- If you're the group leader, absenting yourself from meetings from time to time – your body language can be a dead giveaway
- Considering a team approach, such as dividing into smaller groups to work on the same problem
- Thoroughly scrutinising all the alternatives
- Getting an outsider's perspective or consulting an outside expert
- Choosing one person at random to play devil's advocate at each meeting
- Providing a safe means of reporting if anyone feels the need to escalate an issue

Team size

Team size is extremely important when it comes to agility and balance. Having worked with hundreds of clients, we are firm believers in the Amazon or Jeff Bezos 'two pizza' rule: if two pizzas can't feed the team, it's too big! Communication and interaction between team members becomes significantly more difficult after this point. Projects aside, compare the experience of a dinner party with that of a wedding: as group size grows you can't possibly have meaningful conversations with every person – which is why people start grouping into small clusters to talk.

Small teams make it easier to communicate more effectively, stay decentralised and move quickly. They also encourage a high degree of autonomy and innovation. Large teams require more links between people – it's these links, and the volume of them, that becomes problematic. A team of six has fifteen communication links between members, a team of twelve has sixty-six links, and a team of fifty has an amazing 1,225 links. The cost of attempting to communicate, coordinate and relate to all those people increases exponentially as the team size grows – resulting in decreasing individual and team productivity.

It's a common error to believe adding more people to a problem or project is a good idea. Larger teams tend to breed overconfidence, take longer to deliver and often underestimate effort. Psychologist Jennifer Mueller also notes that large teams are often more stressed. She refers to the loss of productivity with larger teams as 'relational loss': the perception of a lack of support. Adding more people erodes the protective barrier of strong work relationships to mitigate stress and frustration:

> 'In these larger teams, people were lost. They didn't know who to call for help because they didn't know the other members well enough. Even if they did reach out, they didn't feel the other members were as committed to helping or had the time to help. And they couldn't tell their team leaders because [it would look like] they had failed.'

It's difficult to stipulate an exact team size for a project, but Bezos's rule works out to about six or seven people. We advise our clients that anything over ten will likely be unproductive and cause more issues than they resolve. It's worth noting that elite fighting forces, such as the SAS or Navy Seals, tend to work with teams of four. On large projects, such as the deployment of an ERP, you may need more people. If this is the case, break the teams into clearly defined sub-teams with strict boundaries.

When to celebrate

Most projects celebrate significant milestones – design, testing, UAT, etc. While this will set expectations and provide motivation, ensure the celebrated milestones aren't too far apart. The rule is simple: if a piece of work feels difficult at any point in the process, then celebrate its completion. The stronger the team's ethos and bonding, the more cohesive the team – and the faster the project will run.

If you have a large team, then delegate the responsibility of identifying milestones to your team leads – but be sure to invite everyone, including leadership teams. Encourage team socialising as well as project celebration.

How to celebrate

Get creative. People will really appreciate things they haven't done before – give someone a new experience and they will talk about it with others, creating a buzz in the process. New experiences drive enthusiasm and motivation for the next achievement. Take them hang-gliding, rock climbing or parachute jumping. Something incredible. On the flip side, if you take them for a drink, it's unlikely to be memorable or worth talking about with others at work or at home. The more creative you can be, the better – your goal should be to make things as unforgettable as possible. Think of it this way, time is the most valuable resource – it can never be retrieved. Your team are giving up their time for your project, so give them something memorable in return.

Summary

The more experienced and engaged the project team, the faster a project will be delivered. Leaders must put their teams at the forefront of project delivery. All colleagues add value, even those that may appear disruptive. The best leaders get the best from each person on their team – regardless of their role or position. One of the most inspiring examples is from the White House. When Barrack Obama asked a cleaner what his job was, he answered: 'I help run a country.'

The right team, with the right skill levels, is a crucial component of any accelerated project. When the right team is empowered, motivated and rewarded, they will work tirelessly to achieve the goals of the project. The leader must facilitate this, avoiding organisational history and removing obstacles, and allow teams to contribute in the most effective manner. They must communicate frequently and effortlessly – both to internal project members and external stakeholders. They control the motivation, direction, and rails on which the project will run: the rest is up to the team members.

People checklist

- ✔ We have a fully committed executive group and have consulted middle management for project buy-in.

- ✔ Our teams are clear of their day-to-day activities, allowing us to focus on the accelerated project.

- ✔ The *right* people, in terms of capability, are on the project and our team size is optimal for delivery.

- ✔ My project leader is flexible, adaptable and happy with high levels of ambiguity.
- ✔ The leadership teams recognise the importance of removing roadblocks and the organisational change history.
- ✔ We have briefed everyone on how the project will be executed, and they are comfortable with the approach and their level of delegation and autonomy.
- ✔ We have endorsed a self-managed team leadership approach.
- ✔ We have considered a reward network for both project team members and families impacted by the project.
- ✔ We are prepared to help anyone on the project concerned about work–life balance.
- ✔ Our communications (both internal and external) are being continually monitored and updated by someone on the team. We understand the high significance of our communications.
- ✔ We have considered any conflicts or mistakes we may encounter and are suitably prepared to manage them.

- ✔ We know what we are doing to celebration milestones and have some unique ideas to continue inspiring our teams.

CHAPTER EIGHT

Tempo

Setting an aggressive pace at the start of a project is easy: everyone is motivated, charged up and excited about the project. They are likely pleased to be doing something a little different from their normal working day. The unknown is exciting, especially when everyone assumes the project will be easy and run to plan. Some of this naivety likely stems from any sales processes, especially those involving third parties. Salesmen will always make things seem easier than they really are – organisations wouldn't purchase their products otherwise! Executives and leadership teams will also be buoyed by the project, especially if the project is due to deliver in a faster timescale than initially expected. At the beginning, optimism and excitement runs high across the entire organisation.

The challenge comes when the reality and magnitude of the task sets in. Accelerated projects have a tendency to progress like an amateur running a 400m race: they start quickly, underestimating the distance. Then they get tired and begin to slow as the reality of the challenge sets in. But in the closing 50 metres, they find a sudden rush of adrenaline that carries them over the line at speed. Unlike our novice sprinter, a successful accelerated project sets and maintains a consistently fast tempo.

How fast is fast?

A good leader sets the expectations and pace – and then sticks to it. But what is the appropriate pace for a project? It will typically be generated from either a demand–pull or project–push model. In demand–pull, an external deadline is imposed on the project – this might be a new legislative change, a merger or an acquisition. These create a strict endpoint that must be adhered to at all costs. Simply working back from that endpoint and adding in contingency will give you the start point (which may be in the past) and indicates the speed you need to work at to close out your project on time.

The alternative is project–push, wherein the project (or its executives) push and create their own aggressive schedule for delivery. This requires discipline and an acknowledgement of the organisation's capability to execute an accelerated project. The more accelerated projects are run, the more they become embedded within the organisation and culture – and the easier they are to successfully leverage.

In the project–push model, the project lead and senior leaders should determine the size of the challenge they want to give themselves. There are many organisations that take a high-level plan and cut a quarter (or even a third) from it. Others take a more considered and plan-based approach, slicing off time from each project activity. But be warned: when an expectation of timescales being cut is the norm, individuals will develop a natural instinct to pad additional time into their estimates. Encourage realistic planning and challenges: there should be no us-and-them mentality. The project is being accelerated for the benefit of everyone within the organisation.

Encourage teams to plan together using something like 'planning poker' to work out the estimated duration of tasks and the level of acceleration that can be achieved. Planning poker is a concept from the Agile toolkit: everyone is given a pen and card, and each team member is asked to write, without anyone else seeing, how long they think a task, or set of tasks, will take. Cards are turned over one by one and each team member explains their rationale. The group then decides on the estimate together. The project is roughly planned with these estimates, then after a day or so to reflect on discussions, bring the same group together and challenge them to reduce the number. The team are again given pens and cards – but this time write the number they think it could be reduced by given optimal conditions. The net estimate is then incorporated into the project plan.

Maybe the senior team and project leader don't agree on the size of the challenge – after all, why would a project lead want to put themselves under additional pressure? If this is the case, reiterate the why statement to all parties. The pressure is necessary due to the urgency of delivery. Try framing it this way: if a challenge is set to reduce six months from a project schedule, and the project only manages to achieve four, you've still delivered four months early. While we aren't advocating for missed deadlines, we *are* inviting teams to push as hard as realistically possible.

> **TOP TIP:** Accelerate early. Take advantage of the early optimism and get ahead of schedule. The more of a buffer you can create at the beginning, the greater the protection you have against future issues and delays.

Maintaining velocity

Maintaining the velocity of a project once it's underway is a huge challenge: people tire, energy levels drop and the project will inevitably get difficult. These obstacles lead to further issues: motivation drops, mistakes are made, decisions are delayed and people become defensive – ultimately leading to a complete progress slowdown.

This will happen to all projects, accelerated or otherwise – so be proactive. The following are great techniques to maintain productivity.

Motivation and recognition: Ensure your team is motivated to achieve the end goal. The why must resonate throughout the team and everyone needs to be aligned. Use rewards and recognition to boost team morale – it's amazing what a simple pat on the back from someone senior can do for a team member. Take an interest and genuinely show you are pleased and impressed with your team's work.

Gamification: Try to turn aspects of the project into a game. Create a little friendly competition between team members – just be careful to ensure that this is positive competition and seen as fun rather than isolating. A short-term or time-bound competition will help ensure positive and friendly competition.

Remove obstacles quickly: This is where senior managers and leaders can contribute to the project effort. Work with leaders to remove obstacles quickly – this will show your team you're on their side and support them when issues arise.

Involve senior leaders: Show the commitment from the top team – have them unexpectedly buy a round of drinks or a lunchtime pizza. Encourage them to be visible, walking the floor and speaking with everyone as they do so. Ask leaders to reinforce the why.

Challenge sessions: When aspects of the project feel impossible, create small challenge teams. Bringing the team together physically will help solve problems in half the time. Give them the freedom to solve problems as creatively as they like – the only fixed variable is the delivery time.

Rotation: Try mixing up responsibilities or changing work allocations to keep things fresh – a change is a good as a rest!

Fast-track dependent activities: Fast-track activities critical to certain milestones. Start them as early as possible to avoid delaying aspects of the project dependent on their completion. This buys time to keep the team engaged and maintain speed. Never lose sight of these critical tasks.

Do it twice: Aim for speed the first time and accuracy the second. Complete tasks once at speed to create the skeleton, then re-work them with precision. This works great for code writing: get the code in and working, then rehash it for speed, optimisation and all the bells and whistles later.

Team dynamic

An effective leader will consider what is best for each individual member of their team, but also what is best for the team overall – and thus the project itself. Managing the team dynamic involves ensuring morale is high, staff are motivated to complete their individual tasks, and that members are working together as a whole, each completing their tasks at a reliable pace. A chain is only as strong as its weakest link.

Your goal as a leader should be to create a high-performing team. High-performing teams work towards their deadlines and are dedicated to the overall project success. Managing any team to their peak performance level requires identifying under and overperformers: both impact a team negatively and need to be managed appropriately.

Underperformers

An underperformer can spread a lack of commitment through a project if not managed correctly. Unless their performance is addressed, these individuals will demonstrate to the rest of the team that you can get away with slacking off. If not addressed swiftly it will spread to the rest of the team, and an entire team of underperformers is far harder to convert. The following are useful steps to managing an underperformer:

1. Capture metrics to monitor employee performance and spot underperformance early.

2. Give and listen to feedback to find the underlying cause of poor performance. Whatever the reason for underperformance, communication with the employee is essential to the solution.

3. Evaluate the relevant goals and expectations. Is the bar is set too high? Are they overburdened or overwhelmed? Are their personal goals out of line with the project's goals? If necessary, adjust the project expectations, reassign the employee to tasks better suited to their skills, or replace the employee with a more competent and/or dedicated team member.

If your entire team is underperforming, or a subset, then take a collaborative approach. Meet with your team and relevant department heads to discuss the issue and find a solution – but don't vent your frustrations. Keeping an optimist tone will help boost morale and get people motivated and back on track.

Overperformers

There is a big difference between high performance and overperformance. A high-performer is dedicated to the team; an overperformer is motivated by their own ego or career and making team members appear inferior. Overperformers use their high output to abuse company hours, micromanage or wrongfully delegate duties, and deploy demeaning or belittling language. The following are useful steps in dealing with an overperformer:

1. Accept that overperformers are a problem – they aren't an asset to your project. It may be tempting to view this person's work positively and adjust your standards for the rest of the team accordingly. But unless the rest of your team is missing quotas or deadlines, they are the ones working to schedule. Teams work best when everyone moves at an equal or close-to-equal pace. The overperformer is the problem: they are in it for themselves.

2. It's crucial to confront overperformers. If an employee is going above and beyond to the degree that they are making your high-performing members feel inferior, they aren't doing a good job or being a team

player. Overperformers are intimidating, but as the team's leader it's your job to confront them. Discuss you concerns and try to realign them with the company and project. Explain that their contribution is proving detrimental.

3. Rein in overperformers and push them to work for the team. After speaking with an overperformer, assess whether they understand your message on the importance of teamwork and toning down their output to match the pace of their colleagues. If they are looking for more responsibilities or work, assign them the role of team coach. Use the free time they generate from completing tasks quickly to assist and coach other teammates. This will boost morale and teammates will appreciate the added assistance and insights into completing tasks more efficiently.

Know when to let go

All managers know the drain of underperforming staff: missed deadlines and quotas, a demotivated team and a general hindrance on the acceleration of the project. Underperformers stifle the progress made by high-performing team members: it's a simple decision to let them go once you've identified that their work ethic or attitude isn't going to improve.

It's definitely more difficult to release an overperformer from your project. However, Harvard Business School research says that removing a problem employee saves on average $12,500 in team turnover costs – and only 1% of overperformers add more than $5,400 in revenue.[12]

Hopefully your over- or underperformers can be motivated and realigned with your project. Give them an opportunity to improve, but if you don't see real change, don't hesitate to cut ties and save your project team. Remember that good, reliable employees are 54% more likely to quit their job in the presence of a single problem employee.[13]

The cycles of despair

There are several phases a project will move through. These are neatly summed up by Moran and Lennington in their book, *The 12-Week Year.* These phases will affect all team members at various stages, including the leadership team.

Phase 1 – Uninformed Optimism. This is where projects start: sales and business cases are complete and there is firm traction forward. Ideas are flowing, the team is forming and everyone is excited about delivering. There is little to no knowledge of the specific steps required to make it a reality yet.

12 www.hbr.org/2015/12/its-better-to-avoid-a-toxic-employee-than-hire-a-superstar

13 www.cornerstoneondemand.com/sites/default/files/thank-you/file-to-download/csod-wp-toxic-employees

Phase 2 – Informed Pessimism. Here the team starts to understand what it's going to take to make it all happen – it's overwhelming and can start to feel discouraging. The scale, effort, and magnitude are clearly understood, and the team has outlined the requirements for delivery of the project.

Phase 3 – Valley of Despair. The team has started implementing the plan and taking action – and sometimes it feels like the project will never end. The team is intimately aware of the work required and likely struggling with the workload. This is the lowest point emotionally in the cycle – the highest risk of individuals quitting exists at this point in the project.

Phase 4 – Informed Optimism. The team has travelled through the most difficult part of the project, and everyone is starting to see the light at the end of the tunnel. Pieces begin falling into place and a successful outcome feels more real. There is a 'we might just pull this off' mentality.

Phase 5 – Success and Fulfilment. We did it! The project was successfully delivered – and now is the perfect time to reflect. Look at what was achieved, what failed and what's still broken. What have you learned that you will carry forward in your organisation and career?

Particular care should be taken for the team and individuals while the project is moving through the depressive phases. The low emotional state will affect team members and the speed of the project. Giving up isn't an option, so leaders must break the cycle and drive projects into the next phase – and a better emotional state. This takes courage (to carry on, trusting it will lead to success), support (to overcome obstacles and challenges that are slowing the project down), and tenacity (to persevere, even when unsure of the outcome and facing adversity). During low emotional states, bring the project back to its why statement. If the why is greater than the resistance and discouragement, it will (with some reinforcement) keep team members focused on the horizon, rather than their emotional state.

Don't assume the project leader is immune to changes in emotional state. Senior leaders and sponsors should keep in mind the project leader isn't a machine – they too will feel the effects of this cycle. It's the job of senior leadership to refocus the project leader on the why, should they need it.

Burnout

Delivering a project at an accelerated pace is hard work. The project will require seemingly countless actions and tasks in a timeframe that will be a challenge to the best of teams. Leadership will be pushing team members and driving the project aggressively. It's likely team members will be working at close to their maximum capacity over extended periods of time, often working additional hours or through weekends to deliver their part. The very nature of a self-managing team may be stressful to some, as each team is accountable for their own actions. Home pressures will also play a role in stress levels, and without even being aware of it, some team members will reach the end of their operating limit – facing a high chance of burnout.

What is burnout?

Burnout is a serious concern, defined by the Mayo Clinic as: 'A state of emotional, mental, and physical exhaustion caused by excessive and prolonged stress.' It occurs when individuals feel overwhelmed, emotionally drained and unable to meet constant demands. As the stress continues, they begin to lose interest and motivation. Ignored or unaddressed job burnout can have significant health consequences, including:

- Fatigue
- Insomnia
- Strained personal relationships or home life
- Depression
- Anxiety
- Alcohol or substance abuse
- Heart disease

- High cholesterol
- Type 2 diabetes, especially in women
- Stroke
- Obesity
- Vulnerability to illnesses

Burnout may be the result of unrelenting stress, but it isn't the same as too much stress. Stress, by and large, involves *too much:* too many pressures that demand too much of you, both physically and psychologically. Stressed people can still imagine that once they get everything under control, they'll feel better.

Burnout, on the other hand, is about *not enough.* Being burned out means feeling empty, devoid of motivation and beyond caring. People experiencing burnout often don't see any hope of positive change in their situations. If excessive stress is like drowning in responsibilities, burnout is being all dried up. While most people are aware when they're stressed, they won't necessarily recognise burnout.

From a project perspective, leadership must try to protect individuals from burning out, and recognise the signs of burnout early. It must also be a priority to take action to prevent any adverse effects on the person involved – while still working to protect the project and not allow it to stall. Burnout affects the morale of the overall team and dealing with the effects is time-consuming for everyone involved.

Signs of burnout

Burnout is a gradual process. The signs and symptoms are subtle at first but get exponentially worse as time goes on. Acting early can help prevent a major breakdown. Key signs a person is close to burning out are:

Physical

- Feeling tired and drained
- Weak immune system, repeatedly sick
- Frequent headaches or muscle pain
- Change in appetite or sleep habits

Emotional

- A sense of failure and self-doubt
- Feeling helpless, trapped and defeated
- Detachment, feeling alone in the world
- Loss of motivation
- Increasingly cynical and negative outlook
- Decreased satisfaction and sense of accomplishment

Behavioural

- Withdrawal from responsibilities
- Isolation from others
- Procrastinating, taking longer to get things done
- Using food, drugs or alcohol to cope
- Taking out frustrations on others
- Missing work, or coming in late and leaving early

Preventing burnout

Prevention is better than reaction. There are several techniques and tools to prevent burnout from becoming a factor in an accelerated project:

Team Rotation: When roles are able to be rotated around different team members, it can help to release some of the pressure associated with the more stressful ones. In some cases this won't be possible (a finance person can't hand over tasks to a warehouse manager) but where role alignment occurs, look for ways to rotate members through various positions.

Reallocating tasks: Ensure stressful tasks and activities are spread evenly across the team. Self-managing teams should balance their own workloads, but occasionally experienced team members take on too much control. Be sure to monitor this.

Buddy processing: Buddy colleagues up with one another – this enables tasks to be dispersed and balanced, facilitates good social links and communication within the team, and provides coverage in the event of sickness or annual leave.

Social: The social aspects of any project are critical. Allow team members to unwind: arrange disconnect days and ban technology and discussion about work. Talk, have fun and relax with colleagues.

Time off: Encourage everyone to take their holidays and ensure they turn off all work-related notifications and technology while on leave. Plan ahead and use the buddy system to prevent having to contact them – only disturb them in the event of an actual emergency.

Exercise: Encourage team exercise. Some of the best projects strongly encourage regular sporting events – five-a-side football, yoga or running. The more exercise your team gets, the better their health and that of the project.

THE RELAY TEAM

One client implemented a rotational role-switch process for their team members. Each team member could only stay in their role for six weeks at a time before handing over the work to another colleague. The team members stayed engaged and far more connected to their colleagues (and the project in general), chasing each other for updates and trying to stay on top of the latest project information. The handovers weren't easy to manage, but once a process had been identified and the team realised the benefits of this approach, they were keen and transitioned into their new roles quickly.

Dealing with burnout

Dealing with signs of burnout urgently and swiftly is crucial. Encourage anyone suspected of burning out to do the following:

Social contact: Social contact is nature's antidote to stress. Talking face-to-face with a good listener is one of the fastest ways to calm the nervous system and relieve stress. This person doesn't have to provide a solution; they just have to listen to the problem without distraction or judgement. As simple as it sounds, instead of directing your attention to your smartphone, try engaging with your colleagues while on break.

Invest in close relationships: Try to put aside stressors and make the time you spend with loved ones positive and enjoyable.

Limit contact with negative people: Negative-minded people who do nothing but complain will only drag down everyone's mood and outlook. Try to limit the amount of time spent with negative people.

Find deeper meaning: Connecting with a cause or community group that is personally meaningful (like a religious, social, or support group), can give you a place to talk to like-minded people about how to deal with daily stress – and make new friends.

Look for value in tasks: Determine how you help others or provide a much-needed product or service, and focus on aspects of the job that causes enjoyment – even if it's just chatting with your co-workers at lunch.

Strive for balance: Look for meaning and satisfaction elsewhere in life: in family, friends, hobbies or voluntary work. Focus on the parts of your life that bring joy.

Make friends at work: Having strong ties within the workplace can help reduce monotony and counter the effects of burnout. Having friends to talk and joke with during the day can help relieve stress from an unfulfilling or demanding job, improve job performance or simply help get you through a tough day.

Take time off: If burnout seems inevitable, try to take a complete break from work. Use the time away to recharge your batteries and pursue other interests.

Set boundaries: Don't overextend yourself. Learn how to say no to requests when necessary.

Take a daily break from technology: Set a time each day to completely disconnect.

Nourish your creative side: Creativity is a powerful antidote to burnout. Try something new, start a fun project or resume a favourite hobby.

Set aside time to relax: Relaxation techniques such as yoga, meditation and deep breathing activate the body's relaxation response – a state of restfulness that is the opposite of the stress response.

Get plenty of sleep: Lack of sleep can lead to irrational thinking and exacerbate burnout.

Exercise: Exercise is a powerful antidote to stress and burnout. Rhythmic exercise – moving both your arms and legs – is very effective at lifting the mood, increasing energy, sharpening focus and relaxing both the mind and body.

Eat a healthy diet: What you eat has a huge impact on your mood and energy levels throughout the day. Minimise sugar and refined carbs, and reduce the intake of foods that adversely affect mood (caffeine, transfats and foods with chemical preservatives or hormones). Eat more omega-3 fatty acids, as these are proven to give your mood a boost (salmon, herring, mackerel, anchovies, sardines, seaweed, flaxseed and walnuts). Avoid nicotine: smoking when stressed may seem calming, but nicotine is a powerful stimulant that leads to higher, not lower, levels of anxiety. Drink alcohol in moderation: it can temporarily reduce worry, but too much can cause anxiety as it wears off.

Mindfulness in the workplace

In recent years, mindfulness and other stress management interventions have become increasingly popular in the workplace. There is still much debate around the degree of impact these have, and indeed whether they deserve a place in the workplace. The research indicates that mindfulness leads to a reduction in stress and improvements in mood, health, self-efficacy and self-compassion. The following two case studies indicate the positive impact of these processes.

AN OIL COMPANY'S SIXTEEN-WEEK MINDFULNESS TRAINING

A total of thirty executives from a large oil company were offered sixteen weeks of mindfulness training.[14] To test the impact on stress and health in the workplace, researchers collected blood cortisol levels and blood pressure readings, as well as participants' self-reported stress levels and physical and emotional health before and after the sixteen-week programme.

A total of twenty-two executives (twenty-one were male) completed the training, which included a day-long introductory session, and single day training at the end of weeks four and eight. Participants were given audio recordings of daily mindfulness practices and exercises, as well as instructions on coping with stress and a workbook to support their practice. Some executives set up a daily thirty-minute practice group.

At the end of the sixteen weeks, participants reported less perceived stress, improved physical and emotional health, enhanced sleep, better health-related habits and behaviours, and more self-compassion. They also showed significant declines in blood cortisol levels and systolic and diastolic blood pressure, suggesting that both their minds and bodies were less stressed following the programme.

14 Mulla, ZR et al, 'Mindfulness-based stress reduction for executives: Results from a field experiment', *Business Perspectives and Research*, 5/2 (2017) http://journals.sagepub.com/doi/abs/10.1177/2278533717692906

MINDFULNESS AND JOB SATISFACTION

This study looked at whether mindfulness meditation might improve job satisfaction and decrease work-related stress and anxiety.[15] Fifteen faculty and staff from two Australian universities were offered a seven-week modified mindfulness programme. Weekly sixty- to ninety-minute mindfulness lessons were offered for the first four weeks: these included instructions in sitting meditation, performing a body scan and how to integrate mindfulness into daily life. They met again in week seven to revise and refine the participant's skills. Before and after the programme participants completed questionnaires to assess mindfulness, anxiety and job satisfaction. Five members were also selected for interviews after the programme ended.

After seven weeks, employees reported increased mindfulness skill including a heightened awareness of the present moment and physical tension, improved focus, more deliberate actions and less focus on the past and future. Similar to the first study, they also noted improvements in sleep quality.

The interviews provided additional perspective on the benefits of mindfulness for job stress: most notable was a newfound ability to calm down, stay present, and regulate thoughts, feelings, emotions and reactions,

15 Wongtongkam, N et al, 'The influence of a mindfulness-based intervention on job satisfaction and work-related stress and anxiety', *International Journal of Mental Health Promotion*, 19/3 (2017) www.tandfonline.com/doi/abs/10.1080/14623730.2017.1316760?journalCode=rijm20

even in the midst of stressful events. Some participants also reported added personal benefits including more healthy and harmonious relationships with their family, and the ability to leave work behind at the end of the day.

In terms of job satisfaction, overall there were little differences in ratings after the mindfulness programme. There was, however, a significant link between feeling calm and relaxed and greater workplace wellbeing, with those reporting less stress and anxiety also noting higher levels of job satisfaction.

Results of these studies suggest that mindfulness programmes in the workplace may help employees to better deal with stress, and remain calm, present and self-aware, rather than succumbing to negative emotions. Do they work for everyone? Perhaps not. However, it's worth looking into even if it only helps part of your team.

MORE WITH MINDFULNESS

We had the pleasure of working with one client who insisted on mindfulness training for all staff members as they joined the project. They paid for a mindfulness app for each member of the team to have on their personal smartphones which guided and challenged them to complete an exercise each day. Those that completed the exercises for the entire week were entered into a draw to win a meal for their family. The weekly completion rate was around 75%, and there was a noticeable difference to the team: they were far more engaged and grounded, more resilient to the stresses of the project and, on asking, felt mentally healthier as a result of the intervention.

Summary

Starting a project at an accelerated pace is far easier than maintaining its momentum. At the beginning, enthusiasm is abundant and everyone is fresh and keen to get started. As the project continues, constant fine-tuning is required to maintain the initial pace. Team members often have to dig in at this point and work through the challenges of the project, often affecting individual and overall morale and stress. They feel as if the work is endless and the deadlines and deliverables relentless. At this point, leaders need to be highly cognisant of team burnout and manage workloads and stress accordingly. As deadlines pass, the team will gradually see the light at the end of the tunnel – and the pace and morale will naturally pick up as everyone is excited to get to the finish line.

Maintenance of a tempo is probably one of the hardest parts of an accelerated project – and can quickly feel overwhelming. It's important for the entire leadership team to come together and ensure the pace is maintained – not forgetting, of course, the project leads themselves.

While running accelerated projects it's important to recognise the signs of stress and burnout, intervening before it becomes an issue. Techniques like mindfulness training are worth considering as a preventative measure.

Tempo checklist

- ✔ We have set a tempo using either demand–pull (external requirement) or project–push (internal requirement).
- ✔ Leadership understand the timelines and required tempo of the project.
- ✔ We are prepared and have mechanisms to deal with the inevitable drop in productivity.
- ✔ We know how to recognise and deal with under- and overperformers.
- ✔ We are prepared to let people go should the need arise.
- ✔ We have processes in place to deal with the projects when they hit an emotionally low state.

- ✔ We recognise and are alert to colleagues exhibiting signs of burnout and have a strategy in place to manage them.
- ✔ We have considered mindfulness as a mechanism for staying calm, reducing stress and increasing self-awareness.

Conclusion

The business world has changed. The era in which we find ourselves, the digital era, presents a great many challenges and opportunities. Plagued with the mistakes of the past, organisations and their leaders are seeking a structured method to deliver technology projects to meet the needs of this digital era. Expectations have never been higher, the speed of change faster, and cost scrutinised further. This is all true in the context of a highly complex technical environment with three different generations of workers (Baby Boomers, Millennials and Generation Z). As such, projects must have well-orchestrated governance and control, deliver on highly aggressive timelines, and provide a significant return on investment.

I hope this book has sparked a few ideas for your own project delivery strategy. I can appreciate this may present a drastic change for many, but even the consideration and implementation of just some of these components will radically improve your project delivery.

A note of caution: while I endorse using all of the principles in the book, jumping straight into the PROMPT methodology isn't easy, and the journey and tooling needs to be clearly mapped and defined first. The outcome is highly attractive, but the organisation needs to buy into the principles and the journey – and I wouldn't want to set unrealistic expectations. Switching to this ethos mid-project also requires significant consideration to avoid confusing project teams or stifling projects already underway. Think, plan, then transition. Don't just jump!

What next?

I would suggest you reflect on what you have read within the context of your own organisation or project. Where do you feel your strengths and weaknesses are? What would make the biggest impact on your project delivery? What does your journey to accelerated projects look like? To help with this process, we have created a scorecard to help gauge your level of readiness on the PROMPT method. Go to https://runfast.limelightsolutions.co.uk and take the scorecard. It generates a report based on your responses, scoring you on each of the steps within the framework, and indicating where you need to focus first.

I also suggest you download some of the complimentary resources from our website. These contain various workbooks and documents to help you and your team transition to the PROMPT method:

www.limelightsolutions.co.uk/publications/runfast/resources

Finally, I would encourage you to stay in touch – one of the greatest assets in life is collaboration. We love to hear how you are getting on, what your challenges are and how you've succeeded. Let us throw you a few ideas. Making change is easy – making it stick is hard. Keep going and keep improving: after all, as Tony Robbins says: 'Repetition is the mother of all skill.'

Good luck, and stay focused.

Neil

Acknowledgements

No book would be complete without some sincere heartfelt thanks.

I want to thank all those colleagues I have worked with who helped mentor and coach me, and develop my skills. From my early days working in the public sector through to my mentors today, you have all been invaluable to my development. I am deeply grateful for all the opportunities in my career and for those that put their trust in me while I was learning my craft.

Secondly, to all of my clients and customers past and present, thank you. Thank you sincerely for the opportunities to serve. Every one of you has contributed in some way, shape or form to this book. I wish you all the best of luck in your organisations and hope our paths cross again someday.

To my son, Sebastian: I hope one day when you are old enough to understand the contents of this book, you will read it with pride. Putting pen to paper was inspired by our bedtime stories, and you even gave me the title through one of your moments of youthful babble in our kitchen. Daddy loves you.

Lastly to Suzanne, my beautiful wife. Ultimately this is for you. You are the most important person in my life (and Sebastian!). You keep me going no matter how tough the challenge and inspire me on a daily basis. I am a better person because of you. Thank you for everything you do to allow me to succeed. You are an incredible person and a perfect mother. Nothing would be possible without you. I love you.

The Author

Neil has worked in project delivery his entire career, as a technology generalist delivering development and infrastructure projects, and as a specialist in complex global ERP programmes with multimillion-pound budgets. Throughout his tenure he has worked with all project delivery models – Waterfall, Agile, DevOps and supplier methodologies. Neil's projects have spanned numerous industry sectors and he has had the pleasure of working both from an end user perspective and for several of the world's largest consultancies. Neil's successes have earned him a stellar reputation in the delivery space and he is considered a trusted adviser by peers and clients alike.

Neil's vast experience allowed him unique insight into the challenges all companies face before and during project delivery – many of which aren't being addressed with conventional project methodologies. These long-standing challenges are now overlaid with the new challenge presented by the digital era, resulting in one of two outcomes – it can either be catastrophic to an organisation, or act as a springboard to immense success. Neil has capitalised on the nuances of the digital era, added massive value to existing delivery models, and has quickly carved a niche in the delivery of accelerated projects. His team of acceleration wizards work their magic to deliver the seemingly impossible in record breaking time.

Delivering immense project success for his clients through Limelight Solutions, Neil and his team continue to pursue the goal of providing peace of mind project delivery to teams around the globe, allowing their executive teams to sleep easy, no matter how complex the delivery challenge.

Contact details

neil.how@limelightsolutions.co.uk

www.neilhow.com

www.linkedin.com/in/neilhow

@neilhow1

---

www.limelightsolutions.co.uk

www.linkedin.com/company/limelight-solutions

@limelightsol

www.facebook.com/limelightsolutions

Printed in Great Britain
by Amazon